GEOGRAPHICAL READINGS

Applied Coastal Geomorphology

The Geographical Readings series

GEOGRAPHICAL READINGS

Applied Coastal Geomorphology

The Geographical Readings series

Applied Coastal Geomorphology

EDITED BY

J. A. STEERS

Palgrave Macmillan

First published 1971 by
MACMILLAN AND CO LTD
London and Basingstoke
Associated Companies in New York Toronto
Dublin Melbourne Johannesburg and Madras

SBN 333 07282 0 (hard cover)
333 11033 1 (paper cover)

ISBN 978-0-333-07282-0   ISBN 978-1-349-15424-1 (eBook)
DOI 10.1007/978-1-349-15424-1

Contents

List of Plates

Acknowledgements

'The coastline of Djursland: a study in East Danish shoreline development', by Axel Schou, *Geografisk Tidsskrift*, LIX (1960) © Royal Danish Geographical Society

'Wave power and the Djursland coast', by Sofus Christiansen, *Congrès Internat. Géogr. Norden* (1960) © Royal Danish Geographical Society

'The evolution of sandy barrier formations on the East Gippsland coast', by E. C. F. Bird © *Proc. Roy. Soc. Victoria*, LXXIX (1)

'The formation of Dungeness foreland', by W. V. Lewis © *Geogr. J.*, LXXX (1932)

'The coast of Louisiana', by R. J. Russell © Société Belge de Géol., de Paléontol. et d'Hydrol. 1948

'The erosional history of the cliffs around Aberystwyth', by Alan Wood © Liverpool Geol. Soc. 1959

'Land Loss at Holderness, 1852–1952', by Hartmut Valentin © Walter De Gruyter and Co., *Die Erde*, 1954

'The coastal landslips of south-east Devon', by Muriel A. Arber © *Proc. Geol. Assoc.*, LI (1940) 257

'Coastal reefs and islands and catastrophic storms' © D. R. Stoddart 1970

'The east coast floods', by J. A. Steers © Royal Geographical Society 1953

Introduction

THE coast in this country has been for many centuries of interest to those 'that go down to the sea in ships, and occupy their business in great waters', but the scientific study of the coast is a very recent development. Even as a place for holidays the attraction of the coast is relatively new. The first coastal resort in this country – it was then called a spa – was Scarborough. In 1626 a certain Mrs Farrow noticed a medicinal spring in the cliffs, and made it known. She was aided by Dr Wittie, a good propagandist, who also advocated the drinking of sea water because, he claimed, it cured gout and 'all manner of worms'. The habit of drinking sea water for medicinal purposes continued even into Victorian times. But the attraction of the sea for bathing and for holidays gradually increased, and certain places became popular. The association of Brighton with the Prince Regent and of Weymouth with George III are well-known examples. But it was not until the building of railways and the increasing speed and ease of travel that the coast became popular in any general sense.

Even as late as the First World War scientific literature concerning our coast was scarce. There were of course several general books which referred to problems of erosion and accretion, but very few specific works. It is well worth while to glance through the bibliography of D. W. Johnson's *Shore Processes and Shoreline Development*, published in 1919. As far as this country is concerned books like Newell Arber's *The Coast Scenery of North Devon* (1911), curiously omitted from Johnson's list, and W. H. Wheeler's *The Sea Coast* stand out. There were indeed other books both in this country and elsewhere, but they were not numerous. On the other hand there were a large number of papers published in scientific journals, and with these must be included several publications, some in book form, dealing more specifically with harbour construction and related engineering problems. Of outstanding significance were the volumes of the Royal Commission on Coast Erosion (1907–11). The Minutes of Evidence, letters and other detail included in the early volumes are of great interest and should be read by any student of our coast in association with the final, summary volume. It is, I think, true that all these papers and books were of interest to relatively few people apart from those who, in some special or technical field, were concerned with the coast. This was to some extent the result of school and even university curricula. Geology and botany have long attracted distinguished scholars, but relatively seldom did they write on coastal phenomena. Geologists often wrote about the rocks in the cliffs and perhaps about cliff structures; botanists certainly collected and described coastal plants; but geomorphological and ecological studies are almost all comparatively new. Geography at that time was much concerned with capes, bays and rivers, but it completely neglected the study of the ways in which such features are formed. Many schools, even as late as 1914, were primarily concerned with the teaching of classics and mathematics, history and languages and some science; other subjects were taught, but field work and field studies were almost unknown. Anyone who reads, for example, classic school stories of the early part of this century will realise how those 'other' subjects were regarded! That there were always exceptional schools, masters and pupils is fully recognised, but the present generation, with its great emphasis on field studies in the widest sense of the word, will scarcely, if at all, appreciate how much the attitude to nature study has changed even since the end of the Second World War.

There is, I think, little doubt that the better teaching of physical geography, both in schools and universities, has had an enormous influence on coastal studies. This is not the place to enlarge upon this theme, except to point out that in the last two or three decades geography in universities has to an increasing extent included within its scope the scientific study of physiography or geomorphology. Formerly this was regarded as much more a part of geology, and that it certainly is so is a fact, but in university curricula it is now usually much more fully treated in departments of geography than it used to be. Wherever it is studied, I am convinced that in so far as coastal physiography is concerned it should be taught in such a way as to elicit the collaboration of specialists in other fields. Nowadays the inter-disciplinary nature of many subjects is recognised, and the student of coasts can usually seek and find the help of those in disciplines related to his own. This is as it should be, but there is in my view little doubt that the far better teaching of physical geography has led to the present much increased scientific interest in coastal matters. Let us not forget, however, the debt geographers owe to those geologists who were largely responsible for laying the foundations of physiographical teaching in universities.

Douglas Johnson's *Shore Processes and Shoreline Development*, already mentioned, represented a great step forward. For the first time the study of coasts was coordinated and put into a form which at once showed the possibilities of the subject. Until then the young student had had to be content with somewhat general accounts of coast phenomena in textbooks of geology or geography, and unless guided by a good teacher or possessed of great enthusiasm he was unlikely to pursue the subject any farther. Johnson himself was a geologist, in his book he was much concerned with processes, and the analysis of what is happening on a coast – the dynamic effects of waves and currents, of winds and tides. He was also concerned with the origin and evolution of coasts and of particular coastal features; he emphasised the significance of historical studies, and particularly in his later book, *The New England–Acadian Shoreline*, the importance of ecology in relation to coasts is shown to be of paramount interest. Moreover Johnson amassed a bibliography which comprised the work of specialists in other countries and, for the first time, gave us a good overall picture of the development of the study throughout the world.

Today, largely because of the motor-car, the coast is in great demand, not only in this country but in all with a seaboard in a reasonable climate. This pressure implies many problems, and many miles of coast, here and elsewhere, have been spoiled in one way or another. We are not directly concerned with planning, but it is relevant to call attention to the various bodies, national or local, which are concerned with conservation, that is with the right use of the coast. Certain stretches need complete protection; many others can be developed to some extent; some must be wholly given up to urban or industrial use. The National Trust, the Nature Conservancy, the Countryside (National Parks) Commission, numerous county and local trusts and individual owners are all concerned with this problem. The recent very comprehensive regional and general reports of the National Parks Commission[1] give a full analysis of how the coast is used in England and Wales. Scotland is less vulnerable, and so too is the coast of Ireland. But some other countries are under even greater pressure than England and Wales. We do not always appreciate how much of our coast is open to the public.

All this pressure has made an increasing number of people aware not only of problems of car parking and accommodation but also,

[1] The National Parks Commission was reincorporated as the Countryside Commission about the time the regional reports were in preparation. The final reports of the Commission should appear in 1970.

fortunately, of problems of erosion and accretion. It is of vital importance that more and more people should realise something of the natural processes at work on a coastline. It is a fact that on many coasts natural changes take place more quickly than in any other environment, apart from the sudden and catastrophic effects of volcanism or earthquakes or violent river floods. Anyone who takes notice can see the effects of a storm, of a change of wind and wave direction, of the effect of vegetation on mud flats, or of the continual recession of cliffs. What is more, any careful observer can add materially to our knowledge of coastal change. It is at least partly for this reason that an increasing amount of scientific observation and even investigation has taken place in recent years, but even up to the Second World War the volume of work in this country was not great. The Royal Commission's Report did not appear to lead to much active research. Engineers continued to build harbour works and sea defences, but for the most part on what might be called an *ad hoc* basis. Very little quantiative work had been attempted on the movements of beach and offshore materials, and the numerous references to raised beaches and submerged forests, although of considerable interest, did not give satisfying and comprehensive answers to the problems which the field investigations provoked. On the other hand Johnson's influence certainly made itself felt in America, Britain and other countries. Published accounts of parts of a coastline became far more analytical, and careful mapping of spits and other phenomena enabled us to appreciate how features like Dungeness or Cape Cod had evolved. A. G. Ogilvie's careful work on the coasts of the Moray Firth, published in 1923, deserved far greater attention than it received; it was the first attempt in this country to map and analyse a long tract of low coast.

We have been aware for many years of fluctuations of sea-level. That these fluctuations were closely related to raised beaches and submerged forests was fully appreciated, but it was not until 1915 that Daly focused attention on the matter in so far as it concerned coral reefs. It was realised as early as 1865 by Jamieson, in his classic work on Scotland, that there was a well-marked rhythm of coastal movement associated with the ice age. A tremendous amount of work has followed since then, but we are still unable to give a complete interpretation of events. This is partly because it is difficult to measure rise and fall of sea-level, relative to the land, accurately, and partly because it was not always clear precisely what was being measured. Tide gauges are now giving us a great deal of information, but since both land and sea are moving it is not by any means a simple problem to separate their effects. Moreover, exactly how does one measure, relative to the present level of the sea, the height of a raised beach? What part of the beach do you select for this purpose? Loose beach material, thrown up by waves, can give rise to very misleading ideas. A rock-cut bench is seldom if ever level; it usually slopes seawards, it may have a very irregular surface on account of the nature and structure of the rocks forming it, or it may be tilted as a result of land movements. A rapid perusal of many papers will also show that not only are precise measurements difficult to make, but also that in almost every locality there are far too few of them.

It is this lack of precision which detracts from the value of many researches on the coast. I do not in any way underrate what has been done; it was not until quite recently that accurate measurements became possible, and even now 'accuracy' must be regarded with caution, since few if any workers would claim that their conclusions are as precise as they would wish. In this country coastal research was spurred on by two serious floods – that which did so much damage in London in 1928, and the far more extensive storm surge of 1953. The former inspired investigation into flood levels in relation to tidal surges and also into the significance of a downward movement of the land in southeastern England. The way in which this in its turn provoked a reconsideration of the interpretation of the successive geodetic levellings of this country in relation to movements of sea-level was of great interest to everyone interested in coasts. The 1953 flood, which affected a far greater stretch of coast both in these islands and on the continent, made everyone realise something of the problems of defence against sea erosion and flooding. The work subsequently carried out, mainly by the Department of Geography in the University of Nottingham, on the replenishment of Lincolnshire beaches, almost totally displaced in the storm, has been of great significance. Since the Second World War there has been a great advance in the techniques of measurement of displacement of material along beaches and in the offshore zone. We are now in a position to evaluate, quantitatively and often with some exactness, just what is happening on a piece of coast. Previously we had to be content with qualitative assessments. Measurements of this nature must be carried out much more extensively before we can form an overall picture, but in many countries there is undoubtedly a much better appreciation of the problem than was possible even ten or fifteen years ago. Measurements of the vertical accretion of salt marshes have been made for some time and are comparatively easy to do. But we still lack detailed knowledge of cliff erosion. That recession is rapid in boulder-clay cliffs or cliffs of unconsolidated or unresistant materials is clear enough. A good many measurements have been made, for example in Holderness. But seldom has there been any planned scheme for recording loss over a long period of time, and no reliable knowledge is available on the effects of land water, draining out through the cliff face and so leading to slides, as compared with the effects of direct sea erosion. Other items of this nature could be given; the point is that there is great scope for all sorts of quantitative assessment on a coastline, and until we have these we must perforce be content with inadequate data for an understanding of coastal changes.

Allusion has already been made to the inter-disciplinary nature of coastal research. This, to many people, is one of its greatest attractions. The processes at work on a coast are complex, but their general effects are apparent to all observers. Wave action is the most potent factor, but its variations in intensity depend on many factors. Primarily it is governed by meteorological causes, and since these vary greatly in time and place it is difficult to make exact forecasts. Moreover, in many latitudes, wave action is very erratic in its impact. Tidal action is regular and predictable but some of the most devastating storms – surges – follow from the interaction of tidal and meteorological effects, the particular incidence of which is difficult to foresee.

The nature of the coast itself is obviously important. The differences between the east and the west coasts in England and Wales need no emphasis; in the United States there are profounder differences between the cliffed Pacific coast, the low-lying, flat and stoneless coast of the Gulf of Mexico, and the sandy low coast with its many barrier islands facing the Atlantic between southern Florida and New Jersey. Coasts of this type, and those formed by slightly resistant rocks, may change much in the course of time. Flat sandy coasts along which there may be barriers of sand or spits of shingle are very liable to rapid changes. Coasts where there are extensive deltas are subject to subsidence on account of the increasing weight of sediment brought down by the river. For these and similar reasons there is often a close association with archaeology and history. Remains of former human occupation are often found now well below sea-level, and changes in the growth and decline of some coastal towns are often closely related to changes in their harbours brought about by littoral drift. Not only delta coasts, but nearly all coasts, are subject to vertical movements relative to sea-level, caused by isostatic movements of the land masses – nearly always the after-effects of glaciation – and also by eustatic movements of sea-level. The former, like those in delta areas, are vaiiable both in amount and in location; the latter are world-wide, but also mainly the result of the waxing or waning of existing ice sheets and glaciers.

Archaeological remains may allow us to measure the amount of vertical change, and often to date that change within certain limits. This is particularly the case if C^{14} methods can be applied to the finds. But change in a horizontal sense may also be profound. The growth of spits of sand and shingle, and the inundations caused by the blowing of coastal sands, afford numerous examples of the ways in which physiographical processes have played a role, often of great significance, in local or even national history. Historians and others would do well to recognise to the full the climatic changes that have taken place in this and other countries since, for example, the summit of the power of Rome. H. H. Lamb (*Geographical Journal*, Dec 1967) writes:

The history of England, and that of Scotland, has generally been discussed without reference to change of weather or climate except as an occasional intruder whose appearances, pleasant or grim, have been treated as entirely haphazard. The suggestion that our climate has an intelligible history seems foreign, although in Scandinavia it cannot be ignored, and in Iceland the country's history consists of little else. Some claim to have 'proved' that the widespread social changes and unrest in the later Middle Ages were due to economic causes, and therefore not to any change of climate, as if such a thing would not be bound to show itself in the guise of economic stress or dislocation. The object of this study is to show, partly from the physical laws of the atmosphere's behaviour and partly from a wide range of direct observational evidence, that Britain's climate does have a history and that this is something which must concern us. It touches human affairs in many ways and the environment in which we live. And, because of the latest chapters in the story, it poses practical questions that we have to face today — especially in an age of long-term planning.

Anyone who has made a study of coasts will agree whole-heartedly with this! One need only refer to the inundations of sand in the thirteenth and fourteenth centuries in South Wales, to the final covering of the Culbin estate in 1694, to the effects of hurricanes in the Caribbean, the effects of storms on the coasts of the Low Countries and north-west Germany, the Delta plan in Holland and the proposed Thames barrage as illustrations of this theme.

The interrelation of physiography and botany is often extremely well illustrated on the coast. The way in which salt marshes develop as a result of silt and mud gradually accumulating on a sand flat, the gradual colonisation of plants, the upward growth of the marsh surface caused largely by the filtering effect of the vegetation on the silt-laden tidal waters, and the final passage to brackish-water marsh and even dry land, although this last stage is often the work of man. All visitors to a dune coast must be aware of the intricate relationship between sand supply, blown by the winds, and dune growth. The study of cliffs, still in its infancy, demands an understanding of geology, particularly if the structure of the cliffs is in question. The carving in resistant cliffs is often mainly controlled by joints and bedding, but it may be profoundly affected by igneous intrusions and the former history of the land mass in which the cliffs occur. We know next to nothing about the rate of cutting of rocky shore platforms and cliffs of resistant rock. How many of our cliffs are the sole work of the present agents working upon them? Some of the caves in the cliffs of Carboniferous Limestone in the Tenby peninsula in South Wales were formed long before the sea gained access to them; many of the cliffs of Cardigan Bay and elsewhere are but freshly exposed from their cover of boulder clay which spread over them thousands of years ago.

The civil engineer often has a profound influence on the coast. He is largely concerned with sea defences and harbour works, and in building these he has often had a considerable, and not always beneficial, effect on the adjacent parts of the coast. The positions, size and length of groynes and the nature of harbour entrances are two ways in which the littoral drift may be profoundly affected. It is often the case that the engineer has to act vigorously on some local piece of coast in order to prevent erosion. It is not always possible to investigate in advance how his solution will influence the coast on either side. Nowadays, however, problems of this sort are much more amenable to solution. The recent use of tracers – radio-active, fluorescent, or by means of specially marked pebbles – has enabled the engineer to experiment on the volume and direction of beach and offshore drift before building his defences. Even more effective is the use of scale models. The work of this type at the Hydraulics Research Station, Wallingford, has had an admirable influence on the possible effects of coastal defence schemes. It would be rash nowadays if the installation of any extensive groyne system were not preceded by a careful series of tests on a model. So many of the earlier schemes, however well intended, were erected without any experimentation and all too often only in relation to the locality immediately concerned.

It will be apparent that the study of coasts is not only of great interest from a scientific or a historical point of view, but that it is also of vital importance from the economic point of view. Reference was made on a previous page to conservation, and it was suggested that this meant the right use of the coast. At the present time it is difficult to foresee the future requirements of industry. No one is likely to object to the advantages of, for example, the gas supplies from the North Sea or of quicker travel to the continent by means of hovercraft. But there has been no small outcry about how these and other undertakings affect the coast. Amenities and business enterprise are bound to oppose one another in some places, but it is no use being blind to the necessity and effects of new schemes as they affect the coast. There must be increasing liaison on a national scale between all who have an interest in the coast, and all sides must have a fair hearing. It is not easy to find solutions, but reasonable ones are much more likely to be attained if those who can see the coast as a whole, and can give time and thought to the problems in question, are able to take part in discussion from an early stage. More and more care is needed in the development of the coast. Coastal barrages across Morecambe Bay, part of the Solway, the Dee, even the Wash are all freely discussed. Advocates of certain attributes of such schemes make their points frequently and emphatically. The whole matter, in all its aspects, needs discussion on a national scale. A Morecambe Bay barrage may well benefit places like Grange-over-Sands by giving it a permanent water-level; its effect on west Cumberland, if a trunk road is built along the barrage, will be even greater. On the other hand will a Dee barrage necessarily improve the amenities of Anglesey? It would be very interesting to know, as a result of model experiments, what might happen if a barrage were built across the Wash.

Perhaps some readers of this book may at some time in their careers find themselves members of a council or committee that is concerned with the coast. It may be a river board, an urban or rural district council, a harbour authority or a national committee. It is vitally important that members of such bodies should, if possible, have some knowledge of the natural processes that take place on a coast. Problems of rateable value are usually of great local importance, and if taken alone may have a profound and perhaps undesirable effect on the coast. The more the local problem can be seen in terms of the national perspective, the more likely the right solution will be found.

To choose papers for a book of this sort is, to say the least, a rash undertaking, and one that exposes the chooser to an unlimited amount of criticism. There is, however, a companion volume to this – *Introduction to Coastline Development* – and, although the two are distinct and can be used independently, the choice of papers for the two together is of greater significance than the choice for each separately. I have tried to do two things – first to cover the subject (though for obvious reasons not completely so), and secondly to do so in such a way that the papers should be representative of writers of several countries. But a severe limitation is necessarily placed on an editor because it is obviously unwise to compile a book which will be too expensive for those for whom it is primarily intended. This means that the number of reprinted papers must be small and, what is perhaps even more serious, it is out of the question to include long papers or monographs. It would have been advantageous to reprint G. K. Gilbert's 'The Topographic Features of Lake Shores'. Despite the limitation of its title this is a most valuable paper, and every student of coasts should read it, since the analyses and descriptions given by Gilbert apply to features built in the former lakes Bonneville and Lahontan and now exposed so that they can be studied in three dimensions. I also regret that in neither volume has it been possible to include a comprehensive paper on the interrelations of vegetation and physiography in coastal areas. The reader will undoubtedly find other and perhaps more glaring omissions.

In this volume, although all the papers have both a local and a general interest, and all illustrate certain general ideas, the emphasis is focused on their interest and value as studies of particular places.

The sandy and morainic coasts of Denmark afford excellent opportunities for coastal study. Professor Schou and his colleagues, and his predecessor Dr Niels Nielsen, have published some very interesting work on the evolution of the coastline of Denmark. In this paper Schou gives a detailed analysis of a peninsula on the eastern coast of Jutland. Coastal processes are examined in detail, and since Djursland is a small horst, erosion and accumulative phenomena are both represented. The short paper by Sofus Christiansen which follows it analyses the effects of wind and wave power on this peninsula. The two papers need to be considered together.

Dr Eric Bird is an authority on the coasts of Australia. The paper I have chosen for this book deals with one of the great beaches of Victoria. Those who are mainly familiar with beaches in this country are perhaps apt to think that shingle plays an important part in most beaches. This is not so, and the Gippsland beaches are entirely sand. The configuration of the whole area is fully discussed, and Bird explains the formation of the beaches and their relative heights. There are many other beaches of this type around the Australian coast, and if the student wishes to extend his reading to comparable features in America, he will find the reference to Dr Armstrong Price in the bibliography following Professor Schou's paper on Djursland of interest.

In this country we have a magnificent series of coastal features. Dungeness is one of the best known. The account given by Vaughan Lewis is well known and is a valuable contribution to coastal physiography. Few if any other places can offer such a number of shingle ridges, and by careful observation and the use of aerial photographs Lewis has carried their interpretation much farther than have earlier writers. Dungeness, too, is a feature which, with the associated ridges west of the mouth of the Rother, allows a fairly close correlation with history, in the sense that part at least of the evolution of the ridges can be connected with historical events. Many problems, however, remain. It is hoped that at some time a much fuller study will be made of the relations between the shingle foreland and associated ridges and the development of Romney Marsh. It is unfortunate that the power station has been built near the point, but the recent excavations associated with it (see R. W. Hey, *Geological Magazine*, CIV (1967) 361) have shown that the complete explanation of the foreland is still very far from known.

Dr R. J. Russell's paper is not one of his most recent ones, nor is it as comprehensive as some others. On the other hand it gives, in a short and concise form, a clear picture of the evolution of the Mississippi Delta. One of the dangers of textbooks, especially elementary ones, is that their explanations of natural features are often far too simplified or even completely misleading, partly because the space available only allows an inadequate presentation of the features to be described, and nothing for the ways in which they have evolved. This account of one of the great deltas of the world not only illustrates its complexity but also shows the over-simplified nature of many brief accounts. A longer and more detailed account will be found in Bulletin No. 8 of the Department of Conservation, U.S.A. Geological Survey (1936) pp. 3–189, but the paper here printed outlines the general structure of the delta very clearly.

Papers 6, 7 and 8 all deal with cliffs, but cliffs of very different types. Professor Alan Wood has made a careful study of those in Cardigan Bay near Aberystwyth. The Cardigan Bay coast is one that has changed greatly in the last few thousand years. Boulder-clay flats once extended out from the land, but submergence has taken place and the old cliffs are gradually being re-exposed from beneath their coating of boulder clay. In these cliffs there are many bevels and flats which indicate changes of level. Wood's paper is the first to analyse these features. They are by no means easy to explain, but the paper illustrates admirably the point already made in this introduction – the interest which cliff studies offer. It is also to a very large extent a pioneer paper. Dr Hartmuth Valentin deals with a stretch of coast that has suffered more from erosion than perhaps any other in this country. Valentin carried out his work on Holderness when he was resident in Cambridge for over two years. He has made a most careful study of the erosion between 1852 and 1952 on Holderness, and of course also considers that which took place in earlier times. The amount of erosion increases from Bridlington southwards, and Valentin correlates this with the influence of Flamborough Head and the Smithic Bank. He also considers the loss on Holderness in relation to the growth of Spurn Point, a feature more recently investigated by Mr G. de Boer.[1]

The third paper, by Miss Muriel Arber, deals with the best-known coastal landslip in these islands. In her original paper she gave an interpretation which she has subsequently somewhat modified, as outlined in the short appendix which has been included in this book. The whole area of the slip is now a National Nature Reserve. Part of the interest of this slip is the fact that it occurred at a time (1839) when Conybeare and Lyell were able to visit it

[1] 'Spurn Point and its predecessors', *Naturalist* (1963) 113–20; 'Spurn Head: its history and evolution', *Trans. and Papers, I.B.G.*, xxxiv (1964) 71–89; 'Cycles of change at Spurn Head, Yorkshire, England', *Shore and Beach*, xxxv (1967) 13–20.

whilst the features were still fresh. The self-sown ash wood in the great chasm is noteworthy.

The last two papers have something in common between them in that they are largely concerned with the effects of storms, but they deal with very unlike parts of the world. Dr David Stoddart has had great experience in coral-reef areas and is the only man who has been able to map accurately a group of reefs before and after a severe hurricane. His paper has been specially written for this book and has not been published elsewhere. Coral coasts are unlike any others. Reefs of one type or another form barriers and so protect the coast within them. On the other hand there are numerous reefs which are well away from solid land, and some of them bear a cay or sand island. These islands may become fairly stable and often carry a cover of creeping plants or shrubs. In the Caribbean and elsewhere many have been planted with coconuts. In a hurricane, however, major changes are likely to occur, and some may be swept away, especially if they lie in the path of the storm's centre. Stoddart has been able to study these effects at first hand, and his account is of special interest to physiographers and ecologists. The final paper is an account of the nature and effects of the 1953 storm on the east coast of this country. Here, as with the Belize hurricane of 1961, the sea-level was appreciably raised and violent wave action was superimposed on it. Although constant change is taking place on any coast as a result of the normal action of wind, wave and tide, it is necessary to appreciate what may happen in storms. This country was spared the worst effects of the 1953 storm, since the winds were slightly offshore except on the east–west coasts of the Moray Firth, north Norfolk and north Kent. The Low Countries received the full force and suffered far more than we did. Both studies indicate how important it is to appreciate the intensity of a storm, the rate at which the intensity decreases outwards from the centre, the degree of exposure of a particular stretch of coast, and the unusual levels at which wave action is effective as the result of a surge imposed on the tide.

It would be easy to add to the number of papers in this book in such a way that more writers were represented and more examples taken. Some aspects of coasts have been omitted, but that is inevitable in a work of limited size. The primary object has been to draw attention to certain important matters that must be taken into consideration by anyone interested in coastal phenomena. It is not always easy to obtain access to a large number of journals, and even if a particular journal is available, it is seldom that there is more than one copy. This book and its companion volume, whether they are used singly or in combination, should enable the student easily to expand his reading in such a way that he can have at hand examples of the writing of people who have devoted much of their time and energy to coastal studies. It is valuable to have these papers in a handy form, since not only can much be learnt from each one, but a comparison of the styles of the several writers and the way in which they develop their subjects can be made which may well be of help to anyone wishing to attempt research in the subject. All of the papers are accompanied by bibliographies, and these will help the reader to extend his range. One of the papers – that of Valentin – has been translated; nowadays, fortunately, it is not too difficult to obtain translations of papers. Nevertheless it is a great advantage to read papers in the language in which they are written, but this is an asset not enjoyed by everyone. The omission of papers by French, Italian, Spanish or other authors is regretted, but space and cost imposed rigid limitations.

Coastal research is being actively pursued in many countries. The Commission on Coastal Geomorphology of the International Geographical Union has helped greatly in making known to one another those pursuing coastal research in different lands. It is of real value to meet and discuss with experts, and one valuable function of international meetings is that it allows the student to do this, and also to have the opportunity of seeing a part of the coast of another land under expert guidance. The chance of seeing more of our own or any other coast should never be lost. No two places, even a mile apart, are subject to quite the same influences. Even on a straight coast wave action is variable and differs somewhat from one part to another. Moreover it is essential, if possible, to see the coast under different conditions. The effects of even a minor storm may completely alter the configuration of a beach, and there is an enormous disparity between what happens at high and low water, especially on a coast where the tidal range is appreciable. Coastal research offers wide possibilities, and changes, often of considerable magnitude, can and do take place in a relatively short time, a fact which undoubtedly makes the work more attractive to many workers. There should be more than enough in this volume to urge the reader into the field to do some work of his own!

In conclusion it is worth emphasising the great changes that have taken place in the study of physiography since the dominance of

W. M. Davis and his immediate successors. Davis had a profound influence, particularly in America but also in Europe. Today his writings are perhaps of more interest to the historian of physiography than to its practitioners. His explanatory descriptions of landforms threw a new light on the evolution of landscape, but they were based on rigid rules of deduction. He introduced numerous terms and definitions, and almost created a new language. R. J. Russell rightly says in his *River Plains and Sea Coasts* (1967) that his 'elegance in logic commonly outweighed the presentation of sound field observations'. To a limited extent this is true of some of D. W. Johnson's coastal writings. Nevertheless, the student should neglect neither the work of Davis nor that of Johnson; he should in fact study those parts relevant to the subject-matter of this book and note how and why attitudes have changed. W. M. Davis's *The Coral Reef Problem*, never an easy book to read, is nevertheless worth comparison with recent work on coral reefs and islands; D. W. Johnson's *Shore Processes and Shoreline Development* should be studied again after reading recent works on the same subject. The earlier work should in no sense be underrated; the student will learn much from a consideration of the changing points of view.

Any observer today will soon find that coastal problems can seldom, if ever, be fully solved by Davisian methods. The modern student, with his knowledge of statistical and quantitative methods and new techniques, a more detailed understanding of the relative movements of land and sea, a much greater acquaintance, gained through ease of travel, of a wide range of landscapes, and an awareness that tomorrow may well bring forth some new information which will cause him to re-think his problem, realises that there is no simple answer to his research. He must observe and collect as many facts as possible and try to interpret them in the light of experience – his own and that of others. Then he may begin to understand how and why change takes place on a coastline.

1 The Coastline of Djursland: A Study in East Danish Shoreline Development

AXEL SCHOU

ABSTRACT

The peninsula of Djursland on the east coast of Jutland in the centre of Denmark proper has been analysed in relation to shoreline development. All stages of simplification are represented, ranging from initial moraine coasts to totally simplified equilibrium forms.

The geomorphological analysis should be compared with the results of investigations concerning wave force. (See Sofus Christiansen, 'Wave power and the Djursland coast', *Geografisk Tidsskrift*, LIX (1960) and the following paper.)

GEOMORPHOLOGY AND COASTAL DYNAMICS

COAST types as well as shoreline development are results of an interplay between two distinctly different complexes of phenomena, the geological structures of the land area and the marine activities of the surrounding seas. Concerning geological structure, Denmark is an area of glacial accumulation formed by deposition in the varying marginal zones of the Pleistocene ice cap. The terrestrial nuclei which form the skeleton of the Danish landscape pattern consist of the enormous quantities of boulders, gravel, sand and clay, all forming a pattern of moraine landscapes and glacio-fluvial plains.

During the Riss–Saale glaciation the area of Denmark proper was completely covered by an ice cap. In the last glacial period, the Würm–Weichsel glaciation, the extreme limits of the ice cap never extended far enough to cover the southwestern part of the peninsula of Jutland. That it did not do so is to a very large measure the explanation of the great difference between west Denmark and east Denmark today, as to both relief features and coast types. The main stationary line of the last ice sheet through Jutland (Fig. 1.1) is a geomorphological borderline of distinct significance. South-west of this line old moraine landscapes of the Riss–Saale glaciation lie between the vast outwash plains of the last glacial period, and the whole region is characterised by its flat topography. East and north of the line young moraine landscapes are predominant: large-featured hills with steep slopes and great differences in level in the marginal zones alternate with smooth moraine flats and small local outwash plains.

The peninsula of Djursland is situated in the centre of the Denmark-proper area that lies east of the main stationary line (Fig. 1.1); as regards geological substratum and surface layers it is a typical Danish region in which nearly all Danish relief forms are represented.

Concerning dimensions as well as types of marine activity, there are pronounced differences between the North Sea and the inner Danish seas (the Kattegat, the Danish Straits and the Baltic). For example, in the North Sea the maximum height of waves is 5 m., whereas the highest waves observed in the Baltic are only 3 m. The tidal range at the North Sea coast near the Danish–German border is 2 m., in Esbjerg the difference between the tide levels is only 1·5 m., and farther north at the west coast of Jutland this value diminishes. At the Skaw spit near the entrance to the inner Danish seas the tidal range is insignificant. This means that in the Kattegat and the Baltic only a very small tidal wave is generated. The tiny tidal amplitudes which can be calculated here are normally covered entirely by non-periodical level changes mainly caused by wind pressure. Finally, it should be noted that west winds are predominant, as illustrated by the direction resultant of wind work (DRW) calculated for the island of Anholt in the Kattegat (Fig. 1.1).

The different marine environments have created highly differing coast types in west Denmark and east Denmark. This means that the west coast of Jutland is exposed to strong wind and wave activity, the fetch in the north-west direction being more than 1,500 km. as a maximum and 500 km. as a minimum, the depths increasing to 10 m. very near the coast. As glacial deposits offer only slight resistance to wave attack, mature stages in the simplification of the shoreline are reached at all exposed localities. The west coast of Jutland, with its north–south direction, approaches a straight line because of marine activity caused by westerly winds. The direction of the coastline was not determined by the initial relief, but is due to the forces of the sea. The sea has cut cliffs through all earlier hills and built bars in front of the former intermediate bays. An almost uninterrupted zone of dune landscapes has developed along the west coast, continuing southwards on the west coasts of the

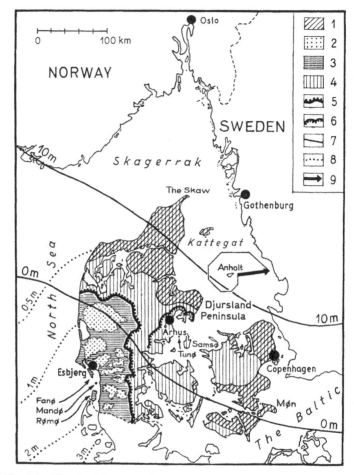

Fig. 1.1   The physical-geographical environment of the Djursland peninsula

1. Area where limestone rocks form the substratum under the Quaternary deposits.
2. Old moraine landscapes, Riss–Saale glaciation.
3. Outwash plains of the Würm–Weichsel glaciation.
4. Predominant young moraine landscapes, Würm–Weichsel glaciation.

5. Main stationary line of the last glaciation.
6. Terminal moraine in the Djursland peninsula.
7. Lines of equal elevation since the Stone Age (Litorina–Tapes epoch).
8. Lines of equal tidal aplitude.
9. Direction resultant of wind work.

islands of Fanø, Mandø and Rømø. Owing to the tides there are salt-marsh coasts along the southwestern part of the shoreline of Jutland.

The east Danish coast type is much more varied than the west Danish type described above. East Denmark being of an archipelagic nature, the fetches are of all dimensions, and the water depth is highly variable, which means that the complex of morphogenetic agencies is to be found in a rich variety of combinations. The resulting form complexes are also influenced by the Post-glacial isostatic and eustatic level changes which are still active. North Denmark is still in a state of emergence, but very slowly, about 1 mm. a year. The southern part of the country has sunk since the Stone Age and is still sinking at a similar rate. It must be added that human activities of many kinds – harbour building, reclamation and coast protection – are factors of great importance in this densely inhabited land.

THE DJURSLAND PENINSULA

The existence of this peninsula on the east coast of Jutland is partly the result of uplift of the limestone substratum between fault lines. Djursland is a horst formation of the bedrock where resistant Cretaceous limestone withstood erosion by the ice sheets of the glacial periods. The shoreline of the north-east corner is bordered by limestone cliffs. The subterranean dislocations which are so characteristic of central Europe were also contributory to the shoreline configuration of Denmark.

Nevertheless Denmark's shoreline is mainly governed by the surface relief of the moraine deposits of the Würm glaciation. Where these accumulations of moraine material are of considerable thickness they form projections on the coastline. This is the case on the south coast of the Djursland peninsula (Fig. 1.2), where the hilly landscapes are explained as marginal moraines formed along the front of a glacier, which during the final Baltic stage of the Würm–Weichsel glaciation, following the Baltic depression, moved from south to north and had its extreme limit here. The bays, Kalvø Vig and Æbeltoft Vig, are submerged central depressions formed by erosion under the ice lobes of this glacier snout.

The peninsula of Djursland is a significant example of various types of Danish moraine coasts as well as of various stages in shoreline development. The north coast is exposed to an open sea

area with a maximum fetch of about 300 km., while the south coast
faces the islands of Samsø and Tunø, north of the Funen archipelago,
where the maximum fetch is not more than 25 km. Governed by these
circumstances, marine erosion and beach drifting have been very
effective along the north coast, which is totally simplified, while the

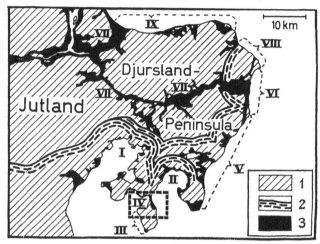

Fig. 1.2   Geomorphological map of the Djursland peninsula

1. Young moraine landscape, Würm–
   Weichsel glaciation.
2. Marginal moraines.
3. Coastal plains formed by marine
   accumulation combined with upheaval of land (3–5 m.) and reclamation. The frame indicates
   the area shown in Fig. 1.3 c.
I. The bay of Kalvø Vig.
II. The bay of Æbeltoft Vig.

III. The totally simplified cliff shore-
   line of Helgenæs.
IV. The bay of Begtrup Vig.
V. The mature simplified festoon-
   shaped east coast.
VI. The totally simplified east coast.
VII. The earlier 'fossil' shorelines of
   the strait of Kolindsund.
VIII. The limestone cliffs.
IX. The old simplified north coast.

southern shoreline still retains all the typical features of the initial
moraine coast.

For coastal research concerning bay closing, tombolo building,
spit growth and the formation of cuspate forelands, a region of this
kind offers the best chances.

With regard to coastal features and stages of shoreline simplification, nine different types may be distinguished in the Djursland
peninsula (Fig. 1.2).

    I. Initial moraine coast – Kalvø Vig, south Djursland.

    II. Young simplification stage of a moraine coast – Æbeltoft Vig,
        south Djursland.

III. The equilibrium moraine cliff shoreline of the west coast of Helgenæs, a beach-drift source locality.

IV. The beach-drift drain locality of Begtrup Vig.

V. Mature, simplified festoon-shaped east Djursland shoreline.

VI. Totally simplified part of the east Djursland shoreline.

VII. Earlier (fossil) shorelines of the Litorina strait of Kolind-sund.

VIII. Tectonically determined limestone cliff coast, north-east Djursland.

IX. Old simplified complex shoreline of north Djursland.

I. *Kalvø Vig, a 'Bodden' coast in initial stage*

The bay of Kalvø, being a submerged central depression, has the dimensions of the ice lobe which generated the initial cavity in the surface relief by glacial erosion. Concerning the shoreline configuration, many details may be explained as results of the landscape-creating activities of the glaciers during the Baltic stage, the last phase of the Würm–Weichsel glaciation. The bay west of the Hestehave woods, south-west of the town of Rønde, is the deepest ('drowned') part of a subglacially eroded valley formed by melt water flowing upwards as a result of hydrostatic high pressure, in a northerly direction in an ice tunnel during the last glaciation. Now this valley contains a small consequent river course running in a southerly direction to the bay. Other indentations of the shoreline, for example Knebel Vig, may be explained as relief cavities originated by resistant dead ice in the Late-glacial period, when the surrounding area was flooded by melt water which caused sedimentation around the ice lumps. The sheltered position of the bay in the angle between the east coast of Jutland and the south coast of Djursland, combined with the narrow inlet, explains the fact that wave activity is only small and the resulting shoreline simplification insignificant. The post-Litorina 2.5-m. land upheaval is responsible for the dead cliff shoreline and the bordering narrow coastal plains, for example on the north coast of Egens Vig. However, even if the wave activity is weak, it has caused small-dimensioned but typical beach-drift phenomena in exposed localities. The island of Kalvø thus was joined to the mainland by a tombolo which was later stabilised by isostatic uplift as well as by human activity – the construction of road and fortifications necessitated by the existence of the medieval castle of Kalvø, of which the tower ruin still remains on the former island. Similar tombolos have developed at Skødshoved, near the

south entrance to the bay, as well as at Dejred Øhoved at the entrance
of Knebel Vig. As a whole the Kalvø Vig shoreline may be characterised as an initial moraine coast only slightly modified by wave
activity and level changes.

II. *Æbeltoft Vig, a 'Bodden' coast modified by shoreline simpli-*
fication

Compared with Kalvø Vig, this bay is more open and exposed to
effective wave attack from the south-east, the fetch in this direction
being 60 km. As a consequence the shoreline simplification caused
by beach drifting has progressed to a certain degree. The lake
Bogens Sø on the west coast is a lagoon lying between the elevated
cliff shoreline of the Litorina sea and the delimiting beach-ridge
plain. In contradiction to Kalvø Vig, the shoreline at the head of
Æbeltoft Vig is not in conformity with the initial relief contours
indicated by the elevated Litorina shoreline. The smoothly rounded
curve may be explained as an approximation to the ideal equilibrium
form of a bay exposed to beach-drift dynamics. Promontories like
Bogens Hoved on the west coast and Ahl Hage on the east coast are
of quite different origin, the former being caused by the resistance of
moraine accumulation, the latter being a cuspate foreland (Plate 2).
The diminishing grain size north along the Ahl Hage beach, in con-
junction with the occurrence of recurved spits on the north coast
of this foreland, demonstrates that in accordance with the general
laws of bay closure beach drift into the bay is responsible for the
formation of this cuspate foreland, which acts as a breakwater
providing the necessary shelter effect for the harbour of Æbeltoft.
The Ahl Hage foreland is based on an extensive submarine sand
accumulation, Sandhagen, clearly shown on aerial photos (Plate 2).
The existence of this extensive accumulation on the eastern shoreline
is governed by the strong beach drift along this shoreline with its
western exposure.

III. *The west coast of Helgenæs, a totally simplified cliff shoreline*

The southernmost prominence of Djursland, Helgenæs peninsula,
does not profit from the sheltered conditions which characterise the
Kalvø Vig region described above. The west coast of Helgenæs in
particular is exposed to effective wave attack generated by the
dominant westerly winds over a sea area with water depths of more
than 20 m. and fetch of about 20 km. As a result of the marine
activity the moraine cliff shoreline from the south cape, Sletterhage

to Stavsøre, at the entrance to the bay of Begtrup, is totally simplified, demonstrating the equilibrium form of the locality carved by erosion. The orientation of the shoreline NNW–SSE is very nearly at a right-angle to the resultant of wave work (Fig. 1.1). This cliff shoreline is retreating without altering its orientation. Small prominences of the shoreline are caused by earthslides, which occur as a result of the retreating of the cliffs after wave attack, in particular in places where lenses of Tertiary plastic clay in the moraine are exposed in the cliff face. The sliding clay masses often exhibit stepped fractures. As a result of these cliff-forming processes, large masses of boulder clay are delivered to the littoral zone and exposed to beach drifting. The coastal stretch is overnourished; in Per Bruun's terminology it is called a source locality of beach-drift material.

IV. *Begtrup Vig, bay closing in different stages*

This bay was a sound in the Litorina age, Helgenæs being at that time an island, which was later joined to the Djursland peninsula by a tombolo (Fig. 1.3c) and afterwards became stabilised partly by the post-Litorina uplift and partly by the building of fortifications on the strategically important tombolo. Exposed to the westerly winds, the beach drift into the bay is very considerable. At the southern entrance the bay acts as a drain locality for the beach drift along the west coast of Helgenæs. Here sand masses have built up a platform on which beach ridges and recurved spit systems are formed at Stavsøre (Fig. 1.3b), the youngest part of a beach-ridge plain constructed in front of the elevated Litorina cliff shoreline. Lagoons in various stages of filling-up by sand accumulation and vegetational growth may be seen (Fig. 1.3b (e), (f)).

The analogous process is to be seen at the northern entrance to the bay (Fig. 1.3a). Here the cliffs of Mols Hoved act as a source of beach-drift material. South of the village of Strands typical stages in bay closure can be demonstrated (Plate 1). At locality a the former bay of the Litorina age is now closed by a bar and the shoreline is completely simplified. Farther to the east, at b, a bay is in the mature stage of closing. A spit complex is in a phase of rapid growth. It is possible to indicate the future locality of bay closure. The aerial photo shows distinctly that enormous masses of sand have been moved into the bay mouth, where they now form the foundation for further spit formation. It is to be foreseen that these submarine accumulation platforms at the north and south

ACG B

shore of Begtrup Vig will combined and form a base on which spit systems from the north and south and from bar islands in the central part may finally be welded together, forming a bar across the entrance of the bay.

V. *The mature simplified, festoon-shaped coast of east Djursland*

The east coast from Hasenøre to Havknude exhibits the festoon shape that is typical of mature stages in shoreline simplification. The prominences Brokøj (34 m.), Jærnhatten (49 m.), Glatved-Limbjerg (40 m.) and Havknude (14 m.) are moraine hills, that at Glatved having a nucleus of limestone boulders. The resistance of these moraine nuclei is a result of the dimensions of the Quaternary accumulations. The festoon parts of the coastline consist of shingle ridges built up between the moraine nuclei as bar islands and tombolos separating lagoons from the sea by the closure of bays and straits between the former moraine islands. Some of these barred parts of the former sea area still exist as lakes, for example Nørresø, east of Rugaard, and Lake Dråby, farther to the south. In other cases the lagoons are overgrown, now forming swampy areas like Gungerne, east of Boeslum, separated from the sea by a beach-ridge plain with a covering of dune sand. The recent beach ridges along this shoreline have a maximum level of 2.5 m. above Danish O.D. Old elevated beach ridges rise to a level of 7.5 m. Earlier coastal features are often truncated by the recent shoreline development, for example at Katholm, where Havknude represents a former island in the Litorina sea. It was separated from the mainland by a strait which today is still identifiable as a low-lying area east of Katholm woods.

VI. *The completely simplified part of the Djursland east coast*

The northern part of the east coast of Djursland, from Havknude to Fornæs, is completely simplified. A beach-ridge plain, Hessel Hede, grew out from the south at the entrance of the Litorina strait of Kolindsund, which is now followed by the course of the river Grenå, the mouth of which has been deflected in a northerly direction by the growth of the spit system. As a consequence of the strong wind activity the Hessel Hede area has been covered by blown sand, this deposit obscuring the structures of the original beach-ridge plain. The conifer plantation established as a shelter against dangerous wind erosion also hides the surface relief. Like many other Danish harbours the original Grenå harbour was localised at the

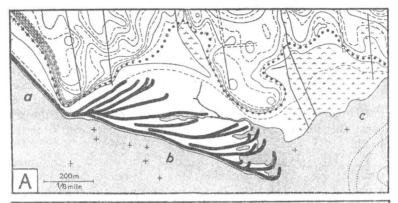

Fig. 1.3   Begtrup Vig: stages in shoreline simplification

A. Bay closing stages. Strands Gunger. *a*. Bay totally closed by coastal plain development combined with upheaval of land (3 m.). *b*. Mature closing stage. Spit complex under growth before the elevated Stone Age cliff shoreline. Surveying 1954. *c*. Former bay, filled up by vegetational growth, sedimentation of fine-grained material and upheaval of land but without beach ridge development until now.

B. Stavsøre. *d*. Northern part of the simplified Helgenæs west coast. *e*. Precipice of the shoreline caused by slides of Tertiary plastic clay. *f. Left*: Old spit complex structures in the coastal plain. *Right*: Recurved spits in the growing stage.

C. Begtrup Vig, localisation map (cf. framed area in map Fig. 1.2).

1. Young moraine landscape. 2. Coastal plain. 3. Cliffs. 4. Shoreline of the Stone Age sea. 5. Dead cliffs.

mouth and the lower course of the river. The need for deeper harbour basins caused by the increasing size of ships has been met by the construction of the modern harbour of Grenå at the sea coast, sheltered by a large pier from the heavy beach drift from the south.

The old stage of simplification, which is indicated by the ruler-straight course of the shoreline stretch described here, may be explained by the fact that calculations of wave force show a maximum value in this locality of the east coast. The adjustment of the shoreline orientation to the terminal direction at a right-angle to the resultant vector of wave force is nearly complete (see Fig. 2.1 of the following paper).

VII. *The earlier ('fossil') shorelines of Kolindsund*

The Litorina strait of Kolindsund mentioned above once used to run eastwards through a depression, originally a subglacial valley which in the Litorina age separated the northern part of Djursland as an island from the mainland. A part of this strait remained as a lake which was reclaimed in the nineteenth century. Its former basin is now a cultivated plain, characterised by the pattern of drainage ditches and the surrounding reclamation canal, running along the dead cliffs of the 'fossil' Litorina sea shoreline. In other places the former strait now remains as swampy areas, the most extensive one being the bay of Pindstrup Mose, from which the peat is used as fuel in the plywood and veneer manufacturing plant which has been built here. Concerning extension of the former strait, see Fig. 1.2.

VIII. *The limestone cliffs*

At Fornæs, near the lighthouse, it is possible at low water to observe a wave-cut platform abraded in solid limestone with a thin veneer of beach deposits. This fact indicates the existence of the underground horst on which the localisation and existence of the Djursland peninsula depend. Farther to the north the Danian limestone rises to a higher level and forms a shoreline of steep cliffs running NW.–SE. The vertical clean-cut cliff walls of Sangstrup Klint and Karleby Klint form the abrupt limit of the land mass of Djursland here at the north-east corner of the peninsula. In the lower parts of the cliffs wave-cut caves are proofs of the force of marine attack. The abrasion plane cut in the limestone is overstrewn with boulders that are being washed out from the fallen parts of moraine masses which cover the glacially smoothed surface of the limestone beds.

In the breaker zone these boulders are eroding potholes in the wave-cut platform, an illustration of the dynamics of abrasion. Surface forms caused by the chemical solvent action of sea water may have some effect too; shallow depressions along the fissure lines of the chalk might be explained in this way.

In Djursland limestone cliffs are only to be seen at this locality, but they are typical elements of the coastal landscapes of Denmark, particularly in the eastern part of the country, where the limestone cliff dimensions are greater than here. This is caused partly by the high position of the Senonian and Danian sediments, partly by the dislocation of the limestone beds by ice pressure in the ice age.

North of the limestone cliff of Karleby Klint a depression forms the limit of this coastal type. The adjacent northern shoreline is characterised by the moraine cliff of Gerrild Klint, and here the typical north-coast type of shoreline starts with a smooth 90° curve, which may be explained as the ideal equilibrium form of coastlines with a finite length, to use the terminology of Per Bruun.

IX. *The old simplified north Djursland shoreline*

In contradiction to the south coast of Djursland, the shoreline of the north coast is quite independent of the initial moraine relief. The main east–west orientation is to some extent tectonically determined, as the underground limestone horst forms the eastern prominence. It is the most simplified stretch of the Djursland coastline, developed partly by the closure of all original bays and sounds, partly by erosion in the moraine deposits. The equilibrium form of the recent shoreline is an approximation to the terminal direction at right-angles to the wave-force vector (see Fig. 2.1 of Paper 2), the great fetch of 300 km. with northern orientation being the dominant factor in the calculation of this value, indicating the strength and orientation of wave action. The smooth curved connection of the northern shoreline of Djursland with the east coast of Jutland may be explained as an adaptation to the equilibrium shoreline of bay-heads, according to Per Bruun's hypothesis.

A geomorphological analysis shows that this simplified shoreline consists of elements of different origins. The western part, Hevring Hede, is a recurved spit complex built out from east to west, closing the northern entrance to the former strait that separated the northern part of Djursland from Jutland during the Litorina transgression. To a certain degree blown sand has obscured the initial structures of this marine foreland, but in many places the fan-shaped pattern of the beach-ridge plain is still indicated by the configuration of the contours in the ordnance sheet, scale 1:20,000. The heavy wind activity of this coastal stretch causes severe wind erosion in the farmlands, in particular in localities with light sandy soils, especially in spring when precipitation is often very small.

The eastern part of the north coast at Knudshoved consists of several moraine nuclei connected by tombolos, in this way separating former sea areas and lagoons, which later became a barred foreland[1] partly by the accumulation of blown sand and partly by vegetational growth.

Like other beach-drift shorelines, the north coast of Djursland is unfavourable to navigation. The need for a fishing harbour has been met by the construction of the Bønnerupstrand harbour of the island-harbour type. This particular harbour construction, with sand-tight moles, surrounds a basin which is connected with the shoreline by a bridge which presents only a minimum of hindrance to the beach drift. This particular Danish harbour type has been adopted with slight modification in many similar localities. Quarrying for gravel and stone, which is an industry of importance, has caused a need for loading facilities, which has been met by constructing extensive piers of sufficient length to reach depths sufficient for the necessary navigation on the broad offshore flat.

There is a significant discrepancy between the recent shoreline and the old mature shorelines of the Litorina age. Elevated Litorina cliff shorelines bordered by coastal plains occur, for example, south of Bønnerup and south-east of Stavnshoved, with an orientation quite different from the shoreline of today. The recent shoreline is thus a complex of accumulation localities and recent cliffs representing an old stage of simplification, the orientation governed partly by the earlier coastline, partly by the recent dynamics.

Even if the Djursland shoreline is of rather limited extent, about 100 km. long, it is possible to find there typical examples of all Danish coast types except the real tidal salt marsh, which to a small degree is represented by some flat shores of the beach-meadow type in sheltered bays where wave activity penetrates far from the outer shoreline. As regards dimensions, the Djursland coastal complex cannot rival certain localities in other parts of Denmark. The limestone cliffs on the island of Møn rise to 100 m. above sea-level, the Rømø beach in south-west Jutland is ten times as extensive

---

[1] A foreland which developed when bars enclosed a bay and made a lagoon which was filled by blown sand on which plants grew.

as any Djursland beach and, compared with the west Jutland dune landscapes, the sand agglomerations on the north coast of Djursland are quite insignificant. Nevertheless Djursland is a typical part of Denmark as regards physical geographical features, and in particular with regard to shoreline development.

REFERENCES

BRUUN, PER (1946–7) 'Forms of equilibrium of coasts with a littoral drift', *Geogr. Tidsskr.*, XLVIII.
—— (1951) (Littoral drift along seashores', *Ingeniøren*, X.
—— (1954) *Coastal stability* (Copenhagen).
CHRISTIANSEN, S. (1958) 'Bølgekraft og kystretning', with a summary: 'Wave power and shoreline orientation', *Geogr. Tidsskr.*, LVII.
GUILCHER, A. (1954) *Morphologie littorale et sous-marine.*
—— *et al.* (1957) 'Les Cordons littoraux de la Rade de Brest', *Extrait du Bull. d'information du Comité Central d'Océanogr. et d'Étude des Côtes*, IX (1).
JESSEN, A. (1920) 'Stenalderhavets Udbredelse i det nordlige Jylland' [The extension of the Stone Age Sea (Tapes–Litorina Sea) in northern Jutland], *D. G. U.*, II Rk, no. 35.
JOHNSON, D. W. (1919) *Shore Processes and Shoreline Development.*
KANNENBERG, E.-G. (1951) *Die Steilufer der Schleswig-Holsteinischen Ostseeküste* (Kiel).
KING, C. A. M. (1959) *Beaches and Coasts.*
KÖSTER, R. (1958) 'Die Küsten der Flensburger Förde', *Schr. des Naturwis. Vereins f. Schleswig-Holstein* (Kiel).
MUNCH-PETERSEN, I. (1918) 'Bølgebevægelse og Materialebevægelse langs Kyster' [Wave movement and beach drifting., *Fysisk Tidsskr.*, XVI.
—— (1933) 'Materialwanderungen längs Meeresküsten ohne Ebbe und Flut', *Hydrologische Konferenz der Baltischen Staaten* (Leningrad).
PRICE, W. A. (1953) *The Classification of Shorelines and Coasts.* Contribution No. 15, Dept of Oceanogr., A. & M. College of Texas.
SCHOU, A. (1945) 'Det marine Forland' [The Marine Foreland], *Folia Geogr. Danica*, IV.
—— (1949a) *Atlas of Denmark I. The Landscapes* (Copenhagen).
—— (1949b) 'Danish coastal cliffs in glacial deposits', *Geogr. Annaler* (Stockholm).
—— (1952) 'Direction-determining influence of the wind on shoreline simplification and coastal dunes', *Proc. XVII Int. Congress.*
—— (1956) 'Die Naturlandschaften Dänemarks', *Geogr. Rundschau* XI.
STEERS, J. A. (1969) *The Coastline of England and Wales* (Cambridge).

2 Wave Power and the Djursland Coast

SOFUS CHRISTIANSEN

ABSTRACT

For some localities of the Djursland coast an attempt is made to estimate the influence of the fetch on wave power. Variations of fetch and coastal material seem to be of major importance for the configuration of the coast, while vertical movements seem to be quite insignificant. For a geomorphological description of the Djursland coast, see Axel Schou, 'The Coastline of Djursland' (previous paper).

THE coastline of the Djursland peninsula in the eastern part of Jutland shows a diversity of morphology which is unusual in so small a region. It is for that reason difficult to establish a standard of reference from which the coast can be morphologically analysed. In this article an attempt is made to use numerical expressions of total wave work fit for specially selected localities. The method by which the expressions are derived was published earlier (Christiansen, 1958), and is based on a formula for effect of wave work $E = W^4HF$, where E means total wave energy from a given direction, W is wind force after the Beaufort scale, H the frequency of wind from the direction considered, and F the fetch in km. The formula was worked out by Per Bruun (1955) and is based on both practical experience and theoretical calculations. From the formula a vector for every compass direction is determined; these are later added geometrically to form a direction resultant. Use of geometrical addition of vectors based on other calculations was earlier made by Musset (1923), Schou (1945), Landsberg (1956) and others.

The wind observations on which the statistics used in this work are based were made between 1879 and 1925 at the Fornæs lighthouse. Of course, the wind varies somewhat in the region, but errors introduced in calculations on this account are considered insignificant, especially as maximum error in observations of wind direction is as much as $22\frac{1}{2}°$. The advantage of using the calculations shown below is therefore not the exactness of the expression of wave work, but the fact that the calculations involve an estimation of the influence of the fetch. As stated already by many workers in coastal matters, variations of fetch at least in closed waters are significant.

The southern part of Djursland, the morphology of which is

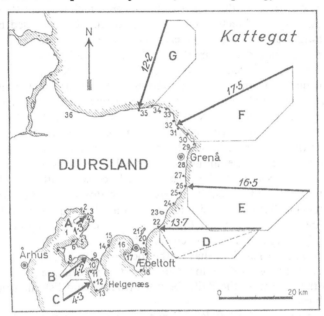

Fig. 2.1   Direction resultants of wave work (DR) along the Djursland coastline
A. Kalvø. B. Strands. C. Ørby. D. Jærnhatten. E. Katholm. F. Gerrild. G. Bønne-rupstrand. 1. Kalvø Vig. 2. Hestehave. 3. Kalvø. 4. Egens Vig. 5. Knebel Vig. 6. Dejred Øhoved. 7. Skødshoved. 8. Mols Hoved. 9. Strands Gunger. 10. Begtrup Vig. 11. Stavsøre. 12. Ørby. 13. Sletterhage. 14. Bogens Hoved. 15. Bogens Sø. 16. Æbeltoft Vig. 17. Ahl Hage. 18. Hasenøre. 19. Brokhøj. 20. Gungerne, Boeslum. 21. Draaby Sø. 22. Jærnhatten. 23. Nørresø, Rugaard. 24. Glatved-Limbjerg. 25. Katholm Sjov. 26. Havknude. 27. Katholm. 28. Hessel Hede. 29. Fornæs. 30. Sangstrup Klint. 31. Karleby Klint. 32. Gerrild Klint. 33. Knudshoved. 34. Stavnshoved. 35. Bønnerupstrand. 36. Hevring Hede.
   Hede=moor. Hoved=cape, point. Klint=cliff. Skov=wood. Sø=lake. Vig.=bay. Ø=island.

mainly glacial, is characterised by the three glacial-depression bays, Kalvø Vig, Begtrup Vig and Æbeltoft Vig (Fig. 2.1).

For *Kalvø*, the direction resultant of wave work (DR) is shown in the illustration. The dimension given in arbitrary units amounts to 1·2. The direction of the DR means that the beach-drifting in the bay has a net movements towards the inner part – as is usually the case for bays. It must therefore be expected that an accretion of material will be found around the small island of Kalvø. It must be noticed, however, that the marine foreland is partly due to the Post-glacial upheaval of the land.

The coastline south of *Strands* has a DR which is larger than that of Kalvø and a direction more to the south. This is caused by the somewhat larger areas of water, which increase the work of waves. The accumulation of material is for that reason larger than for Kalvø Vig. As a sign of accumulation, there is a very nice complex recurved spit which is of some use as a natural breakwater for fishing dinghies.

Measured in the same units, the wave power of *Ørby* is a little larger. The resistance to wave attack of this area has been so small that the glacial forms of the initial coastline are hardly recognisable. In fact the coastline is almost linear, and it will most likely maintain this form. Wave power along this coast will show but slight variation, and the recession of the coastline will consequently be almost constant per unit of length. There is little doubt that the coast in this locality has developed into one of the plane-equilibrium forms described by Per Bruun (1946). It will be noticed that in accordance with the more violent movement of coast material, its average grain size is larger than that at the localities previously mentioned. (Some of the pebbles are indicator boulders revealing a glacial drift from the bottom of the Baltic between Sweden and Estonia (quartz-porphyries).)

The form of *Æbeltoft Vig* is probably a constant but in this case recurved. Per Bruun (1946-7) postulated a special recurved plane-equilibrium form. The postulate has met strong criticism. If lines indicating the dominant fetch are drawn from points along the beach, they are most often orthogonal to the coast. This means that such points of beach may probably be regarded as points on a curve of equilibrium. If the inlet of the bay is narrow, there is a tendency for it to develop a semicircular coastline. Probably this is so in Lulworth Cove, Dorset, and also at Æbeltoft Vig. The latter inlet seems, however, to be too wide to make the explanation valid; it is in fact greatly narrowed by an area of shallow water which excludes all waves of large magnitude.

The east coast of Djursland differs *in toto* from the coasts already treated. *Jærnhatten* (i.e. the 'Iron Hat', an old-fashioned helmet) is a morainic cliff developed by strong wave power. The large fetch from the east results in a DR from that direction. A small truncated foreland is situated in front of the cliff. The form of this can be explained by splitting the DR in to two components. These will be orthogonals to each of the two sides of the foreland. The sides of the foreland can then be regarded as equilibrium coasts. Of course the splitting is only allowable if the two sides are quite distinct; this can be the case if the bottom indicates a diversion in this place. In fact the problem of cuspate forelands (the 'Dungeness problem') is still not definitely solved.

The magnitude of DR by *Havknude* and especially by the cliffs of *Sangstrup* and *Gerrild* shows a maximum for the region treated in this paper. In spite of this the coastline around the points mentioned is far from the straight-line form developed at Ørby. The cause of this is a resistance to abrasion on the east coast which considerably exceeds that of the average moraine coast. At Jærnhatten the resistance was conditioned by the large amount of wave work required to move the masses of material concentrated in the cliff and on the bottom in front of it. Farther to the north the cliffs do not consist solely of glacial deposits, but are based on *limsten* (a limestone belonging to the Cretaceous system). Because of the varying resistance to erosion, the east coast of Djursland has developed the characteristic 'festoon' form.

The north coast has no reinforcements of limestone and is for the most part built up by accumulations of beach material. This is why the coastline by *Bønnerup*, in spite of its smaller DR values, is straighter than the east coast. On account of beach-drifting – as indicated by the DR from the east – a small harbour by Bønnerup is built according to the 'island' principle.

Calculations of wave power on the north coast are made difficult by the variations of the fetch. Fetch of significance is found in only a few directions, but in those cases they are far longer (300 km.) than is usual in enclosed waters. The difficulty of such long fetches is, that the data of observations do not show if an observed wind force is accompanied by the expected maximum height of waves (or of wave energy). With a small fetch the problem is not difficult, because of the observation frequency, which is four hours. In four hours lesser wind forces will, over a small fetch, be able to raise maximum wave height. Evidently this cannot be expected when the fetch is 300 km. Wave power at Bønnerup must therefore, even on a conservative estimate, be about 1,200–1,500 'units'. Incidentally one is puzzled by the relatively frequent winds of force 10 in the wind-observation tables. The explanation must be the human tendency to prefer easy numbers!

Four factors seem to determine the development of coasts: (1) the initial form; (2) the structure of the coast (mass of material, kind of material and grain size; (3) wave power; and (4) vertical movements of coast.

The importance of the initial form is clearly seen by comparing the coastline of Strands with that of Ørby. Though identical in most respects, their initial forms make them very different; the former is a young coast, and the other has reached maturity.

The effect of difference in structure is easily seen from the fact that the eastern coasts of Djursland have not yet reached maturity, as has the Ørby coast. Limestone seems to exert a resistance against abrasion at least three to four times as great as boulder clay. By comparing the two coasts it must, however, be noticed that they differ in length. The length of a stretch of mature coast can under certain conditions be regarded as an expression of its stage in development – or, what is almost the same, an expression of the available wave power.

Vertical movements seem to be of no importance in the region concerned. Measurements show that the north coast is rising and the south coast sinking. When differences in initial forms are considered, this movement cannot be deduced from the morphology. Effects of very slow vertical movements on Danish coasts are much less significant than the work of waves on the loose deposits. For this reason systematics of Danish coasts cannot be based on the Davis–Johnson system; the view of Gulliver seems more appropriate.

REFERENCES

BRUUN, PER (1946–7) 'Forms of equilibrium of coasts with a littoral drift', *Geogr. Tidsskr.*, XLVIII.
—— (1955) *Coast Stability* (Copenhagen).
CHRISTIANSEN, S. (1958) 'Bølgekraft og kystretning', with an English summary, *Geogr. Tidsskr.*, LVII.
DANISH METEOROLOGICAL INSTITUTE (1933) *Danmarks Klima: Climatic Record of Denmark* (Copenhagen).
LANDSBERG, S. Y. (1956) 'The orientation of dunes in Britain and Denmark in relation to wind, *Geogr. J.*, CXXII (2).
SCHOU, A. (1945) 'Det marine Forland', *Folia Geogr. Danica*, IV; *Medd. f. Skall.-Lab.*, IX.

3 The Evolution of Sandy Barrier Formations on the East Gippsland Coast

E. C. F. BIRD

ABSTRACT

Two stages of development are recognised in the sandy barriers of the east Gippsland coast; a late Pleistocene stage (the prior barrier and parts of the inner barrier in the Gippsland Lakes region), separated by features of dissection and rearrangement during the Last-glacial phase of low sea-level from the Recent stage (the rest of the inner barrier, together with the outer barrier), added during and since the Post-glacial marine transgression.

INTRODUCTION

THE coast of East Gippsland (Fig. 3.1) is bordered by a series of sandy barrier formations. An *outer barrier*, extending for the whole length of the Ninety-Mile Beach, is backed by a narrow tract of lagoons and swamps, and then a line of bluffs, facing seawards, which mark a former cliffed coastline at the margin of a gently undulating plateau of Tertiary and Pleistocene rocks. Between Letts Beach and Red Bluff, the former cliffed coastline recedes behind an embayment of intricate configuration, the East Gippsland embayment, which has been sealed off by barriers to form the Gippsland Lakes (Bird, 1965). Here, in addition to the outer barrier, there is an *inner barrier* enclosing Lake Wellington, Lake Victoria and Lake King, and parts of a *prior barrier*, so called because it originated at the head of the East Gippsland embayment before the Gippsland Lakes were enclosed by the inner and outer barriers.

It is suggested that the prior barrier, together with parts of the inner barrier, formed during a late Pleistocene phase when the sea stood at, or a few feet above, its present level; that these barriers were dissected by stream incision and partially rearranged by wind action during a subsequent low sea-level phase, evidently the Last-glacial phase; and that the outer barrier, together with part of the

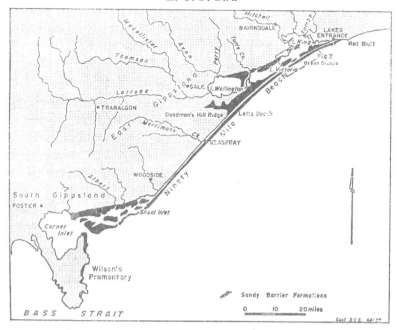

Fig. 3.1 Barrier formations on the East Gippsland coast

inner barrier, developed when the sea returned to its present general level in Recent times.

The evidence for this chronology comes mainly from the Paynesville district, in the Gippsland Lakes region (Fig. 3.2), where the prior barrier is represented by Banksia peninsula and Raymond Island, the inner barrier by Sperm Whale Head and the Boole Boole peninsula (including Jubilee Head), and the outer barrier by the dune ridges at Ocean Grange and the Ninety Mile Beach. These form the visible surface tracts of a large mass of generally unconsolidated and mainly sandy sediment, banked upon a coastal ledge of a consolidated Upper Tertiary, and possibly Lower Pleistocene, rock formations in a manner previously described and illustrated (Bird, 1963, p. 236). Similar masses of Quaternary deposits have accumulated on many parts of the Australian coast as beach and barrier formations, often enclosing lagoons or tracts of swamp land (Thom, 1965; Langford-Smith and Thom, 1965), and it is clear that the East Gippsland barriers represent at least two phases of deposition as the sea-level rose and fell during Pleistocene and Recent times.

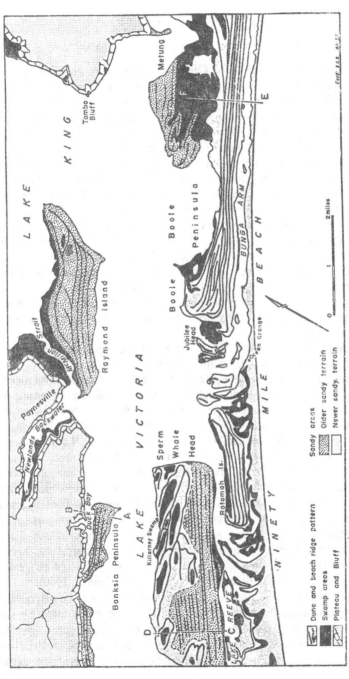

Fig. 3.2 Barrier formations in the Paynesville district (Boole Boole Peninsula refers to the whole of the inner barrier west from Metung to Jubilee Head)

THE BUILDING OF COASTAL BARRIERS

Sandy barriers of the East Gippsland type were formerly called offshore bars (e.g. Johnson, 1919), but modern workers have followed Shepard's (1952) suggestion that the term 'bar' should be restricted to features submerged by the sea for at least part of the tidal cycle, and that depositional forms above normal high-tide level should be termed 'barriers'. Johnson (1919) deduced that wave action would build a barrier parallel to the coastline in the offshore zone if the depth of water were suddenly reduced by emergence, due to uplift of the coast or a fall in sea-level. Alternatively, a barrier may originate as a spit, elongated parallel to the general outline of the coast, or across the mouth of an estuary or embayment. Both mechanisms of formation require the delivery of sediment to the developing barrier, either eroded or collected from the sea floor and carried shorewards (onshore drifting) or brought along the shore from either direction (longshore drifting), and most barriers have been nourished by a combination of the two processes. Further reference to the problem of barrier initiation will be made after considering the evidence from East Gippsland.

Sandy barriers on the Australian coast are typically surmounted by beach ridges (berms) built parallel to the shoreline by wave action. These have evidently formed successively, as a consequence of the alternation of 'cut' and 'fill' on a shoreline prograding by sand accretion. The beach profile is 'cut' during stormy weather, when short, steep waves scour away the sand, whereas 'fill' takes place during calm weather, when long, low ocean swell delivers sand to the shore and builds up a berm along the length of a beach (Davies, 1957). Parallel foredunes may be added when colonising vegetation traps wind-blown sand on beach-ridge foundations (Bird, 1960). Successions of parallel beach ridges, with or without surmounting dunes, commemorate the former alignments of a prograding sandy shore, alignments which have evidently been determined largely by the dominant pattern of constructive ocean swell approaching through coastal waters, and refracted to gently curved outlines by contact with the sea floor (Davies, 1960).

The pattern of parallel beach ridges and dunes may be rearranged during subsequent cycles of 'cut' and 'fill', or interrupted by the development of 'blow-outs', which often grow into larger, migrating parabolic dunes, with advancing noses of spilling sand and trailing arms held in place by vegetation. Blow-outs and parabolic dunes

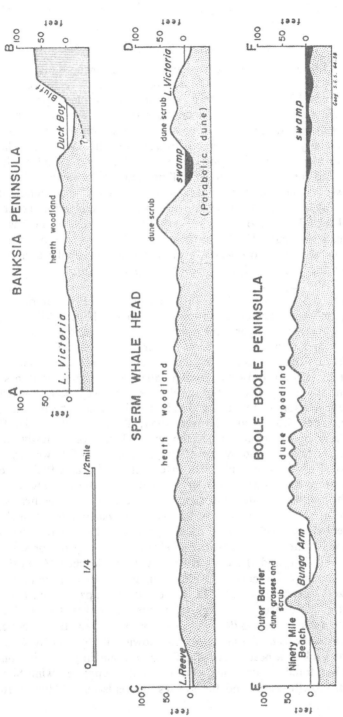

Fig. 3.3 Transverse profiles of barrier formations shown on Fig. 3.2. AB, prior barrier at Banksia Peninsula. CD, inner barrier at Sperm Whale Head. EF, outer barrier near Grange Ocean and inner barrier at Boole Boole Peninsula.

formed in this way have axes aligned with the onshore resultants of wind action, as determined from directional wind-vector diagrams (Jennings, 1957). Their pattern interrupts and displaces pre-existing parallel beach ridges and dunes. Blow-outs are generally initiated where the retentive cover of dune vegetation is damaged or destroyed, particularly where a foredune is truncated at the back of the shore by storm-wave action, laying bare a cliff of crumbling sand in which the wind carves out a hollow, spilling some of the sand landwards. Blow-outs may also form where the edge of a vegetated dune is cut back by river or tidal scour, or where the vegetation cover is weakened by fire, overgrazing, aridity or excessive trampling by animals or man. If the initiating factor, whether erosional or ecological, ceases to operate, a blow-out or a parabolic dune may be arrested and stabilised by recolonising vegetation.

On the East Gippsland coast the barriers consist mainly of quartz sand, with only a small proportion of shelly material. The proportion of carbonates in the sand on the Ninety Mile Beach rarely exceeds 10 per cent, and it is probable that the beach ridges and dunes of the barriers here were built from similar material. The characteristic succession of soil and vegetation features seen on transects across parallel beach ridges and dunes of quartzose sand leads to the development of deep podzol profiles and vegetation communities dominated by heath species on the oldest sites (Burges and Drover, 1953; Turner *et al.*, 1961). The leaching of carbonates from the upper layers of a newly built beach bridge or dune by percolating rain water is followed by the removal of the iron oxides that give fresh sand grains their yellow colouring, and organic matter derived from dune vegetation is washed down through the sand and accumulates, together with some of the leached iron oxides, as an illuvial horizon of lightly cemented sandrock ('coffee rock'), generally close to the level of seasonal water-table fluctuations. This is essentially the process of podzolisation, and the end-product, a deeply leached 'A' horizon over an illuvial sandrock 'B' horizon, is termed a 'ground-water podzol' (Stephens, 1962). The process is accompanied by, and to some extent depends on, a vegetation succession that starts with the grasses (chiefly *Festuca littoralis, Spinifex hirsutus*, and the introduced *Ammophila arenaria*) that trap wind-blown sand to build foredunes at the back of a beach. Growing foredunes remain grassy, but once a newer foredune develops, cutting off the supply of wind-blown sand, growth ceases and the grasses are replaced by 'dune scrub' communities, dominated by *Leptospermum laevigatum*, with the coastal banksia tree *Banksia integrifolia* common. Under dune scrub the sand is leached to depths of 2–3 ft, the surface sand having a *p*H value of 5·5 to 6·5 (compared with about 8·0 for fresh dune sand), with no shell material remaining. On the older beach ridges and dunes, scrub is replaced by 'dune woodland', with *Eucalyptus viminalis* the dominant tree, and an undergrowth of bracken (*Pteridium esculentum*). Here, the dune sand may be leached to depths of more than 10 ft, the leached zone being underlain by a layer stained brown by the accumulation of organic matter and iron oxides, but not yet a firm coffee rock. Surface sand has *p*H values in the range 5·0 to 6·5 and, as acidity increases, the coastal banksia gives place to the saw banksia, *Banksia serrata*. This tree shares dominance with *E. viminalis* on the oldest beach ridges and dunes, where 'heath woodland' is developed, the bracken under-growth giving place to communities of heath shrubs (e.g. *Epacris impressa, Hibbertia acicularis, Astroloma humifusum, Amperea xiphoclada*, and the localised dotted heath-myrtle, *Thryptomene miqueliana*, abundant on Sperm Whale Head). Locally, the heath is almost treeless. The soils are profoundly leached and strongly acid (*p*H considerably below 4·0 at the surface), with a firm coffee-rock horizon at depth.

The succession of soil and vegetation features across parallel beach ridges and dunes clearly represents an age sequence from the youngest, newly developed on a prograding shore, to the oldest, towards the landward margin. Where blow-outs and parabolic dunes have developed, the transverse age sequence of soil and vegetation features has been interrupted.

THE PRIOR BARRIER

The prior barrier, traceable from the north side of Lake Wellington eastwards to Banksia peninsula and Raymond Island, originally developed in front of a cliffed coastline at the head of the East Gippsland embayment. The probable configuration at this stage is shown in Fig. 3.4A, with Tom's Creek flowing into a lagoon behind the western half of the prior barrier, and the outlet from Newlands Backwater deflected northeastwards, through McMillan Strait, to open into what is now the northern part of Lake King. This recon-struction is based on the pattern of beach ridges which form the ground-plan of the dissected remnants of the prior barrier. Best preserved on Banksia peninsula, these run roughly parallel to the bluffs that lie behind them. Their crests are spaced at intervals of 100–150 yds, and their amplitude (from crest to swale) is 5–15 ft. The swales stand generally 5–10 ft above the calm-weather level of Lake Victoria, the lake level being approximately equivalent to mean sea-level in Bass Strait. In section (Fig. 3.3A–B), the topography is subdued and there is a slight seaward fall. The sand has been thoroughly leached, and a well-defined coffee-rock layer is found about 5 ft beneath the swales and up to 10 ft beneath the crests; it is horizontal or gently undulating and a little above calm-weather lake level. Heath woodland vegetation is dominant.

These features of morphology, soil and vegetation suggest that the prior barrier is of considerable antiquity, and the slightly elevated swales may indicate that sea-level at the time of prior barrier formation stood a little higher than it does now. But there has been much subsequent dissection. South of Paynesville, a broad strait has been cut through the prior barrier, separating Raymond Island from Banksia peninsula, and wide embayments have been formed intersecting the barrier on the north shore of Lake Victoria. In addition, there are other outgrowths in the form of recurved spits and cuspate forelands, the largest of which has grown southwards to separate Lake Wellington from Lake Victoria. Parts of the sandy terrain lying north of these lakes are therefore of comparatively recent origin; several of the cuspate forelands on the north shore of Lake Victoria are still being enlarged by sand accretion. The shores of Banksia peninsula and Raymond Island also show marginal depositional features of more recent origin added to the dissected remnants of the original prior barrier.

THE INNER BARRIER

The inner barrier is of composite origin and has had a complex history. The ground-plan of beach ridges and dune ridges at its southwestern end, south of Lake Wellington, shows that it originated as a recurved spit which was prolonged intermittently northeastwards across the mouth of the embayment, and afterwards widened by the addition of successive parallel beach ridges and low foredunes on the seaward side (Bird, 1963, Fig. 3.1). The parallel ridges are well preserved on Sperm Whale Head (Fig. 3.2), where their dimensions and spacing (Fig. 3.3C–D) are similar to those on Banksia peninsula. The swales are again 5–10 ft above calm-weather lake

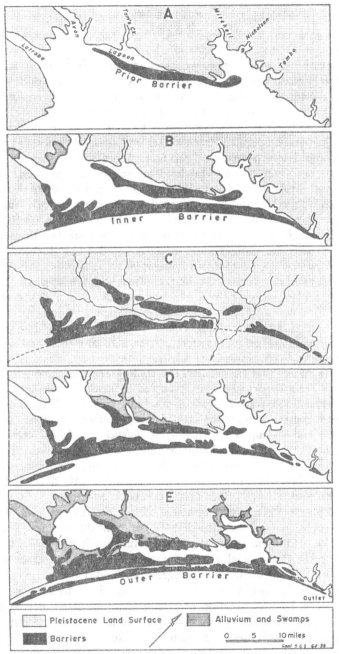

Fig. 3.4 Evolution of barrier formations in the Gippsland Lakes region

level, and there are deep podzol profiles, with coffee rock at depth, and a heath woodland vegetation, all suggestive of an age comparable with that of the prior barrier. It appears that the recurved spit, which became the first inner barrier, grew across the mouth of the embayment soon after the prior barrier had formed, enclosing a lagoon system on the site of the present Gippsland Lakes (Fig. 3.4B). The inner barrier has been much modified since it first formed. The pattern of parallel beach ridges and dunes has been partially rearranged into a group of parabolic dunes which migrated eastwards until they became stabilised in their present position. These are well displayed on Sperm Whale Head (Fig. 3.2), where the older beach ridges, low and widely spaced, are in sharp contrast with the adjacent parabolic dunes which are locally more than 90 ft high. Soil profiles on the parabolic dunes are only a few feet deep compared with the ground-water podzols on the undisturbed beach ridges, and the *Leptospermum laevigatum* scrub and dune woodland on the parabolic dunes is in sharp contrast with the heath woodland on the older beach ridges. The evidence of soils and vegetation therefore confirms the idea that rearrangement into parabolic dunes took place after the original formation of low and widely spaced beach ridges on the inner barrier, and came to an end when a vegetation cover became re-established on the parabolic dunes.

The swales that lie between the trailing arms of these parabolic dunes are now swampy flats, occupied by lagoons after heavy rain or when flood waters invade from Lake Victoria. They generally have a central unvegetated clay plain, surrounded by zones of salt marsh and swamp scrub (mainly *Melaleuca ericifolia*) vegetation; Killarney Swamp, on Sperm Whale Head, shows these typical features. Borings have shown that the swamp deposits, chiefly organic clays and silts, extend to a depth of at least 20 ft here, and so the deflated sandy floor of the parabolic dune must lie considerably below present level and well below present sea-level. From this evidence it is concluded that the parabolic dunes developed at a time when the sea stood at a lower level than it does now, that is before the Post-glacial marine transgression brought the sea to its present general level. This transgression flooded the swale to form a narrow lagoon, which has since been filled to calm-weather lake level by accumulation of swamp deposits. It is inferred that, during the Last-glacial phase of low sea-level, late in Pleistocene times, the basin now occupied by Lake Wellington and Lake Victoria drained out, and that the Latrobe and Avon rivers, together with other tributaries, extended their courses along the emerged furrow and found a way out across the sea floor to the low sea-level. The broad gap in the inner barrier east of Sperm Whale Head probably marks the site where river drainage escaped at this stage. The prior barrier was now breached by the outflow from Tom's Creek, which then became incised into a lacustrine plain which is the emerged floor of the lagoon mentioned previously in the section on the prior barrier, and from Forge Creek (Newlands Backwater) which found a southward outlet between Banksia peninsula and Raymond Island. Lake King must also have drained out, so that the Mitchell, Nicholson and Tambo rivers flowed across its floor and out through the broad gap in the inner barrier. The configuration at this stage is shown in Fig. 3.4c. The inner side of the inner barrier was probably cut back, with initiation of parabolic dunes, as the result of scour by the extended Latrobe river flowing along what is now the floor of Lake Victoria.

During the succeeding marine transgression, the rising sea flooded back into the lagoon basins and the outer barrier was built up as an additional seaward rampart enclosing the Gippsland Lakes. The swales of parabolic dunes on Sperm Whale Head were flooded and, as swamps began to develop in them, the dunes, no longer activated by wind-drifted sand, became colonised by vegetation and stabilised in their present outlines. By this time, however, an additional change had occurred in the inner barrier, for the eastern half (the northern portion of Boole Boole peninsula) stands at a lower elevation than the western half, Sperm Whale Head. Low, widely spaced parallel ridges, similar to those on Sperm Whale Head, are still traceable on the northern portion of Boole Boole peninsula and the soil and vegetation features are also similar, but the intervening swales stand at or below present calm-weather lake level, compared with an elevation of 5–10 ft on Sperm Whale Head. The swales are occupied by tracts of swamp land, bordered by salt marsh and swamp scrub vegetation (Bird, 1962), but eastwards (south of Metung) they widen and coalesce as the beach ridges vanish beneath a broad swamp. Traced by probing, they remain widely spaced and parallel (Fig. 3.3EF, right-hand portion). This evidence suggests that the inner barrier has been tilted transversely, either by tectonic elevation of the section farther west, or because of subsidence of Boole Boole peninsula, which could result from compaction of underlying deposits, such as compressible peats, interbedded at depth. The immediate consequence is that the younger foredunes of the outer barrier, which farther west developed a mile or so seaward of the inner barrier shoreline, have been built *on to* the southern part of Boole Boole peninsula (Fig. 3.4D); although geographically part of the inner barrier, these foredunes are undoubtedly of Recent origin, having developed as the Post-glacial transgression submerged the southern margin of the older barrier. There is a succession of high (20–30 ft) and closely spaced (30–50 yds) parallel dunes (Fig. 3.3E–F), with dune woodland on soils that are leached to about 10 ft, but with no development of true coffee rock. The parallel dunes extend west to Jubilee Head, a complex recurved spit on the eastern side of a 'tidal delta' of shoals, low islands and channels, which marks a gap in the barrier system open to the sea until very recently. On the western side of this gap there is a matching recurved spit at Rotomah Island, and it is probable that, at one stage, this section of the barrier extended farther to the south-west. The southern margin of Boole Boole peninsula has an extremely fresh appearance and was clearly a beach, open to the sea, until the outer barrier developed in front of it.

THE OUTER BARRIER

Westwards from Sperm Whale Head, the outer barrier stands about a mile seaward of the south shore of the inner barrier, separated from it by Lake Reeve, a tract of sand flats, salt marshes and shallow lagoons. The Ninety Mile Beach forms its seaward margin and it is surmounted by a series of high (20–80 ft) and closely spaced (30–50 yds) parallel dunes. At Ocean Grange there are two of these, but westwards the number increases to a maximum of thirteen at Letts Beach, where the vegetation on parallel dunes shows the early stages in succession from grasses on the unleached sand of newly built foredunes to dune scrub and woodland on the moderately leached sand of the inner ridges. Depth of leaching increases from dune crest to dune crest on landward transects away from the Ninety Mile Beach, confirming that these dunes were built successively on a sandy shore that has prograded. The oldest, on the landward side, has been leached to a depth of about 5 ft, but although the underlying sand is stained reddish-brown, the accumulation of down-washed organic matter and iron oxides has not yet reached the status of coffee rock. The absence of heath vegetation and the rarity of the saw banksia (*B. serrata*) on the dunes of the outer barrier support the idea that these parallel dunes are of no great age.

The initiation of the outer barrier is not easily explained, for it does not conform exactly with either of the usual explanations of barrier formation mentioned previously. As the south shore of the inner barrier at Sperm Whale Head shows no sign of any recent modification by the waves of the open sea, the outer barrier must have come into existence during the later stages of the Post-glacial marine transgression as the sea approached its present general level. The pattern of parallel dunes yields a little more evidence, for the multiple foredunes at Letts Beach indicate a section of the barrier that developed at an early stage as a barrier island. Traced laterally, the inner dunes curve away successively to recurved terminations in Lake Reeve, the number of parallel dunes diminishing in this manner northeastwards and southwestwards from Letts Beach. The barrier island initiated offshore at Letts Beach was evidently elongated northeastwards and southwestwards and prograded to take up the present alignment of the Ninety Mile Beach. South-westwards, the growth of the outer barrier cut off a formerly cliffed coastline at Seaspray, and continued to its present termination as a recurved spit south of Woodside, with Shoal Inlet on the inner flank. Growth to the north-east was irregular, for there is a series of curved channels leading from Lake Reeve into the back of the outer barrier west of Rotomah Island (Fig. 3.2), which testify to the former existence of gaps in the barrier, diverted northeastwards by longshore drifting before they were finally sealed off. The features are similar to those described by Lucke (1934) from coastal barriers in the vicinity of Barnegat Inlet, New Jersey, where gaps in the barriers have migrated and closed under the influence of longshore drifting. Current action in these deflected channels evidently trun-cated the southwestward extensions of Rotomah Island, fragments of which are traceable in the sandy terrain on the south side of Lake Reeve. The continued growth of the outer barrier north-eastwards eventually outflanked the gap between Rotomah Island and Jubilee Head, Bunga Arm persisting as an outlet channel that also suffered northeastward diversion and final closure. The story was completed by the extension of the outer barrier as far as Red Bluff, the Cunninghame Arm at Lakes Entrance being the last of the deflected outlets from the Gippsland Lakes, still active at the time of discovery in 1839 (Fig. 3.4E). The cutting of an artificial entrance through the outer barrier at Lakes Entrance in 1889 stabilised the outlet from the Gippsland Lakes and led to the sealing of the former natural outlet farther east (Bird, 1961).

The initiation of the outer barrier in the vicinity of Letts Beach may have been a consequence of an exceptionally abundant near-shore sand supply on this section of the coast, or it may have been prompted by localised tectonic uplift of the land. It is perhaps significant that Letts Beach lies upon the seaward continuation of the Deadman's Hill ridge, an anticlinal area of Tertiary rocks in a region that has been subject to tectonic deformation during Quaternary times (Boutakoff, 1955); its uplift could still have been in progress when the Post-glacial marine transgression came to an end and, indeed, may have been part of the movement which gave the inner barrier a lateral tilt. Elongation and progradation of the barrier island initiated here resulted from the continued delivery of large quantities of sand to the coast, and sand has been spread along the shore in either direction by the action of waves and associated currents.

In these terms, it is not necessary to invoke an episode of general 'Recent emergence' to explain the initiation of the outer barrier, although the possibility of localised emergence due to uplift of the land has been mentioned. The view that the Post-glacial marine transgression rose to a higher level about 4,000–6,000 years ago and then dropped back to its present stand has been widely accepted by Australian coastal geomorphologists following Fairbridge (1948) and others, but it has been criticised, notably by Shepard (1961) and Russell (1963), on the grounds that the evidence is not world-wide in the manner required for a eustatic oscillation of sea-level. The widely reported evidence of Recent emergence on the Australian coast may result from uplift of certain sections of the coast late in Quaternary times; on the East Gippsland coast, Recent emergence, if it has occurred at all, has evidently been localised in a manner suggestive of tectonic uplift.

PRESENT-DAY SHORELINE EROSION

The long-continued progradation of the sandy shoreline of East Gippsland appears to have come to an end at least temporarily, for during the last few decades 'cut' has exceeded 'fill' along the Ninety Mile Beach and new foredunes have not developed. Instead, the outer edge of the youngest dunes has been truncated by wave attack and blow-outs have been initiated, with sand spilling landwards across the outer barrier. North-east of Ocean Grange this erosion may soon breach the outer barrier and reopen Bunga Arm as a natural outlet from the Gippsland Lakes.

Evidence of a very recent phase of shoreline erosion is widespread on the sandy shores of southeastern Australia, and has been attributed to a renewed eustatic rise of sea-level, perhaps accompanied by increasing storminess in coastal waters (Davies, 1957). An alternative suggestion is that the erosion is due to a reduction in sand supply in coastal waters, the consequent steepening of the offshore profile allowing more powerful wave action to attack the shore (Langford-Smith and Thom, 1965), but this reduction is probably associated with either or both of the factors noted by Davies. The absence of cliffing on the foredunes preserved behind the sandy forelands that have developed since 1889 alongside the protruding stone jetties at Lakes Entrance places the onset of erosion within the past century (Bird, 1960), and within the period for which world-wide tide-gauge analyses suggest a secular eustatic rise of sea-level (Valentin, 1952).

THE SOURCE OF THE SAND

The traditional explanation of the origin of the East Gippsland barriers is that sand has been swept northeastwards along the coast from the vicinity of Wilson's Promontory by a powerful ocean current (Gregory, 1903), and that this was a sequel to the foundering of the 'land bridge' that formerly extended across Bass Strait between Tasmania and the mainland (Hall, 1914), but these hypotheses cannot be accepted in the light of modern knowledge of coastal evolution. The 'powerful ocean current' does not exist, and the weak ebb-and-flow tidal currents which occur off the Ninety Mile Beach cannot have moved much sand, but sand is transported northeastwards along the Ninety Mile Beach as the result of longshore drifting by waves and associated currents generated when strong winds drive in waves from the south-west, and in the opposite direction when the waves come in from an easterly direction. As the westerly winds are prevalent, the drift to the north-east probably exceeds that to the south-west. The balance is a fine one, however, for the similar scale of beach accumulation on either side of the protruding stone jetties at Lakes Entrance indicates that similar quantities of sand have arrived here from both directions (Bird, 1961).

There is little evidence that a major source of sand existed formerly in the vicinity of Wilson's Promontory and, as the plunging coastal slopes of resistant granite have not been cliffed by marine erosion since the sea attained its present level, they cannot have produced large quantities of sand. In any case, the barriers have not grown northeastwards from Wilson's Promontory, for the prior and inner barriers originated in the East Gippsland embayment about 50 miles north-east of the promontory, and the outer barrier terminates in a recurved spit which has grown southwestwards *towards* Wilson's Promontory. The suggestion that barrier formation was linked with the making of Bass Strait is ruled out by the evidence that Bass Strait has existed intermittently since late Tertiary times during the rise and fall of Pleistocene eustatic oscillations of sea-level (Jennings, 1959*a*); it was finally revived by the Post-glacial marine transgression, and was already in existence when the sea rose towards the East Gippsland coast during the later stages of that transgression. More generally, there seems no need to link the formation of the East Gippsland barriers with the origin of Bass Strait, since there are similar barriers on the New South Wales coast, and in Encounter Bay on the South Australian coast, which can bear no relation to the formation of straits or to any such changes in adjacent coastal configuration.

The origin of the East Gippsland barriers has evidently depended more on the onshore drifting of sand and the effects of refracted ocean swell in supplying sediment and determining shoreline alignments than on longshore drifting of coastal sand. Longshore drifting has played a part in the growth and shaping of the barriers, but the bulk of the material has been eroded or collected from the sea floor and carried shorewards. It is most obvious in relation to the outer barrier, which was prograded by sand accretion even after it had developed in front of the cliffed coasts and river mouths which might otherwise be regarded as possible sources of sand for barrier construction. The sand evidently came from deposits that were previously laid down on the sea floor as barriers or dunes when the sea withdrew to a low level during the Last-glacial phase. There is now little evidence of these depositional forms, for the sea floor off the Ninety Mile Beach was smoothed over by wave action during the succeeding marine transgression, when sand was collected and carried shoreward to build and nourish the outer barrier, but, off Flinders Island, Jennings (1959*a*) has identified submarine ridges of uncertain origin which could be relics of submerged barrier or dune formations that have survived destruction by the rising sea. The formation of barriers off the present shoreline during a low sea-level phase, and their subsequent destruction by the waves of a transgressing sea to provide sand for the building of newer barriers, has also been postulated by Hails (1964) to explain certain features of beaches and barriers on the New South Wales coast. The concept of landward sweeping of sea-floor sediments during the Post-glacial marine transgression helps to explain many aspects of Australian coastal beach and barrier formations.

CONCLUSION

The chronology presented turns on the recognition that certain features of rearrangement and dissection of the prior and inner barriers in the Gippsland Lakes region originated when the sea stood at a lower level, preceding the Post-glacial marine transgression. This was evidently the Last-glacial phase, when sea-level is believed to have fallen at least 300 ft (Shepard, 1961). The original formation of the prior and inner barriers is therefore placed back in a late Pleistocene interglacial (or interstadial) phase when the sea stood at or slightly above its present level, and the addition of the outer barrier took place in Recent times, during and after the Post-glacial marine transgression. Certain features of the barriers suggest the influence of Quaternary tectonic deformation of this section of coast: the evidence of transverse tilting since the original formation of the inner barrier, and the possibility of uplift as a means of initiation of a section of the outer barrier as a distinct feature offshore in the vicinity of Letts Beach.

In terms of this chronology, it is suggested that barriers which have been built of quartzose sand on the coasts of southeastern Australia are likely to be of Pleistocene origin where they show deep podzolic profiles with true coffee rock at depth and a heath or heath woodland vegetation, but of Recent origin where they show evidence of incipient podzolisation without true coffee rock and a vegetation of grasses, scrub or dune woodland without heath communities. This is essentially the distinction made by Jennings (1959b) in recognising Old Dunes of late Pleistocene age and New Dunes of Recent age on the coasts of King Island. The distinction is less clear where the parent sand material is strongly calcareous, as on the west coast of King Island, the west side of Wilson's Promontory, and much of Australia's south and west coast, where a higher base status reduces the rate of podzol formation and succession to heath vegetation on dunes and beach ridges lithified as calcarenites. On the other hand, if the quartzose sands are extremely poor in

62   *E. C. F. Bird*

original shell content, as on parts of the coast of Wilson's Promontory, dunes and beach ridges of Recent origin may show the advanced podzolic profiles and associated heath communities that are elsewhere typical of older coastal sand deposits.

REFERENCES

BIRD, E. C. F. (1960) 'The formation of sand beach ridges', *Aust. J. Sci.*, XXII 349–50.
—— (1961) 'Landform changes at Lakes Entrance', *Vict. Nat.*, LXXVIII 137–46.
—— (1962) 'The swamp paper-bark', *ibid.*, LXXIX 72–81.
—— (1963) 'The physiography of the Gippsland Lakes, Australia', *Zeitschrift für Geomorph.*, VII 233–45.
—— (1965) *A Geomorphological Study of the Gippsland Lakes* (Australian National University, Canberra).
BOUTAKOFF, N. (1955) 'A new approach to petroleum geology and oil possibilities in Gippsland', *Mining & Geol. J. (Vict.)*, V 39–57.
BURGES, A., and DROVER, D. P. (1953) 'The rate of podzol development in the sands of the Woy Woy district, New South Wales', *Aust. J. Botany*, I 83–94.
DAVIES, J. L. (1957) 'The importance of cut and fill in the development of sand beach ridges', *Aust. J. Sci.*, XX 105–11.
—— (1960) 'Beach alignment in southern Australia', *Aust. Geogr.*, VIII 42–3.
FAIRBRIDGE, R. W. (1948) 'The geology and geomorphology of Point Peron, Western Australia', *J. Roy. Soc. W. Aust.*, XXXIV 35–72.
—— (1961) 'Eustatic changes in sea-level', *Physics and Chemistry of the Earth*, IV 99–185.
GREGORY, J. W. (1903) *The Geography of Victoria* (Melbourne).
HAILS, J. R. (1964) 'A reappraisal of the nature and occurrence of heavy mineral deposits along parts of the East Australian coast', *Aust. J. Sci.*, XXVII 22–3.
HALL, T. S. (1914) 'Some notes on the Gippsland Lakes', *Vict. Nat.*, XXXI 31–5.
JENNINGS, J. N. (1957) 'On the orientation of parabolic or U-dunes', *Geogr. J.*, CXXIII 474–80.
—— (1959a) 'The submarine topography of Bass Strait', *Proc. Roy. Soc. Vict.*, LXXI 49–72.
—— (1959b) 'The coastal geomorphology of King Island, Bass Strait, in relation to changes in the relative level of land and sea', *Records of the Queen Victoria Museum, Launceston*, XI 1–39.
JOHNSON, D. W. (1919) *Shore Processes and Shoreline Development* (New York).
LANGFORD-SMITH, T., and THOM, B. G. (1965) 'New South Wales coastal morphology', in *Geology of New South Wales*.
LUCKE, J. B. (1934) 'A study of Barnegat Inlet, New Jersey, and related shoreline phenomena', *Shore and Beach* II 1–54.
RUSSELL, R. J. (1963) 'Recent recession of tropical cliffy coasts', *Science* CXXXIX 9–13.
SHEPARD, F. P. (1952) 'Revised nomenclature for coastal depositional features', *Bull. Amer. Ass. Petrol. Geol.*, XXXVI 1902–12.
—— (1961) 'Sea-level rise during the past 20,000 years' *Zeitschrift für Geomorph.* suppl. 3 (*Pacific Island Terraces: Eustatic?*, ed. R. J. Russell) 30–5.
STEPHENS, C. G. (1962) *A Manual of Australian Soils*, 3rd ed. (Melbourne).

THOM, B. G. (1965) 'Late Quaternary sand deposits between Newcastle and Seal Rocks, N.S.W.', *Proc. Roy. Soc. N.S.W.*, XCVIII 23–36.

TURNER, J. S., CARR, S. G. M., and BIRD, E. C. F. (1961) 'The dune succession at Corner Inlet, Victoria', *Proc. Roy. Soc. Vict.*, LXXV 17-33.

VALENTIN, H. (1952) 'Die Küsten der Erde', *Petermans Geog. Mitt. Ergänzungsheft*, 246.

4 The Formation of Dungeness Foreland

W. V. LEWIS

DUNGENESS FORELAND comprises an area of reclaimed marshes and shingle wastes symmetrically placed between the headlands of Fairlight on the west and Shorncliffe on the east. The Ness (Fig. 4.1) extends in a southeasterly direction to the 10-fathom contour, and from this point the shores run to the west and north respectively for about fifteen miles, curving round to join the lines of the old cliffs. The shingle is mostly concentrated in three masses, the main one to the south and east of Lydd and two lesser ones to the east of Winchelsea and the south of Hythe respectively. The general trend of the ridges comprising these shingle forelands, as taken from Sheet 4 of the One-Inch Geological Survey Maps, is represented in the figure. A hummocky strip of mixed sand and gravel, distinctly above the level of the marshes on either side, runs from Lydd to New Romney and then northeastwards as the Warren, to join the present shoreline south of Dymchurch.

Dungeness has been studied closely in the past on account of its importance as a topographical feature, but no satisfactory explanation of its formation seems to have been given. An early theory which attracted much attention (Appach 1868, p. 16) at the time was due to Elliot (1847), an engineer of the Romney Marsh. He invoked the presence, on the site of New Romney, of an island of Hastings Sands, to form a nucleus round which further supplies of shingle and sand collected; but later Elliot realised that these sands were entirely wind-blown in origin, marking the site of an ancient shoreline.

Drew (1875, p. 16) seems to have been the only authority who studied closely the action of the sea on the foreshore in an attempt to elucidate the problem. He gave an excellent description of the way in which the waves threw up the shingle into ridges at high-water mark, and explained how the lessening of the rate of the drift eastwards immediately to the lee of the Ness would lead to the deposition of ridges along that section of the shore; but Drew seems wrong in his attempt to explain its original formation. He considered that banks of sand were first deposited as a result of the

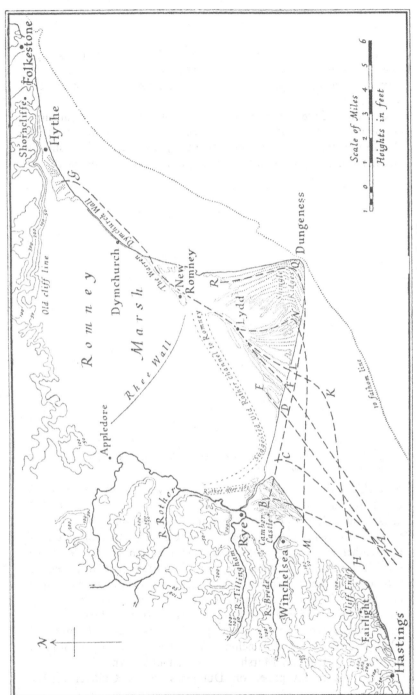

Fig. 4.1 Dungeness, showing general trend of shingle ridges

meeting of the Channel and North Sea tides, and that the shingle was later thrown up on these sandbanks. The tides, however, have been shown to meet over a much broader zone, varying between Dungeness and the Goodwins.

One of the fullest attempts to account for the origin of Dungeness is due to Redman (1852–3). He refuted the suggestion of deposition due to the meeting of the tides, citing the parallel case of the shingle foreland of Langney Point, near Eastbourne, and explaining that

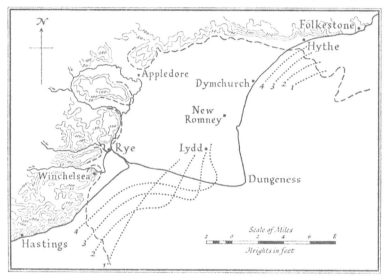

Fig. 4.2 Original coastal outline and successive positions of the 'Dungeness' shorelines (after Gulliver, 1897)

both this and Dungeness could not occupy the slack water at the meeting-point of the tides. The origin of the whole foreland he left unexplained, but considered that the area extending from Lydd to the Ness was built from the shingle held up by the old mouth of the Rother at New Romney. This accumulation of shingle to the windward of an estuary has played an important part in the earlier formation of the foreland, but the Ness cannot be due to this action because it still grows eastwards though the Rother has deserted its Romney channel for nearly seven centuries. In the *Memoirs of the Geological Survey*, Topley (1885, pp. 211, 304) recounts much of what Drew wrote, but concludes that 'The cause of the original formation of Romney Marsh is altogether unknown'.

The most recent paper on Dungeness is by Gulliver (1897).

He attributes cuspate forelands to tidal and eddy currents, and applies these views to Dungeness in spite of the weight of evidence to the contrary. It is fully recognised in this country (Royal Commission on Coast Erosion, 1911), and by the more recent authorities in America (Tarr, 1898; Woodman, 1899), that the velocity of tidal currents in actual contact with the foreshore is quite inadequate to drift the coarse material which comprises the shingle forelands (Lewis, 1931); and further, so far as the writer is aware, there is no evidence of the existence of such large-scale eddy currents. Although there seems little doubt that his explanation is inaccurate, it is suggestive in giving an ordered series of changes in the outline of the foreland.

Gulliver starts his present cycle of shore erosion after the submergence which caused the drowned valleys around our coasts, and considered that the outline of the land at that time would be produced by extending the surface slopes above the cliffs till they reach sea-level. In Fig. 4.2, taken from his paper, the dotted line represents this outline and the broken lines 1, 2, 3 and 4 are successive positions of the 'Dungeness' shorelines. The first, 1, is a simple bay bar from headland to headland, but as these promontories suffer erosion, the bay bar changes gradually to a cuspate foreland. Apart from disagreement with his tidal explanation of the cuspate form, it seems clear from the existence of the old cliffline on the north and west of the marshes that the whole feature is of more recent origin than this cliff. That is, there have not been nearly five to seven miles of erosion at Fairlight and Shorncliffe headlands, as Gulliver suggests, since Dungeness began to form. According to the estimate by de Lamblardie (1789), with which Wheeler (1902, p. 200) also agrees, such chalk cliffs recede at the rate of about 1 ft per annum, thus requiring on Gulliver's hypothesis a period of over thirty thousand years for the formation of Dungeness, an estimate which seems excessive considering that it has passed through a great part of its evolution in the last 1,500 years. In his book *The Sea Coast*, Wheeler (p. 198) also attributes the Ness to tidal eddies distributing the shingle held up by an earlier mouth of the Rother. It seems natural therefore that in a recent reference to the subject Ward (1922, p. 206) repeats the conclusion reached by Topley, that a satisfactory explanation of the formation of Dungeness has not been forthcoming.

Osborne White (1928, p. 82) again appeals to the interplay of tidal currents producing an area of relatively slack water favourable to the deposition of shingle, and quotes Gulliver's paper at some length as offering the best explanation. Attention should be drawn to this apparently official recognition by English geologists of what seems to be an obsolete tidal-current hypothesis.

In his account of the Dymchurch Wall, Elliot (1847, p. 466) suggests that at an early date there was a natural barrier of shingle to the seaward of the site of this wall which formed continuous ridges from New Romney to Hythe. Recurves are visible along this stretch of coast running into the marshes, and in his opinion they were formed when an ancient shingle spit grew towards Hythe. On approaching Hythe the longitudinal extension became slower, so that isolated recurves gave place to lateral ridges forming a continuous shingle foreland. Elliot attributed the recurves to the influx of the tides at the old Rother estuary which, according to certain records, was once situated at Hythe. This hypothetical coastline consisted of continuous ridges of shingle running from Fairlight Head through point *F*, Lydd and New Romney (see Fig. 4.1), and then seaward of the present shoreline to Hythe.

Lewin (1862, p. 55), Montague Burrows (1888, p. 10), Drew (1875, p. 308), and A. J. Burrows (1884–5) follow Elliot in suggesting this even line of coast from Fairlight to Hythe as an early stage in the evolution of Dungeness. They presumed that during the extension of this shoreline there was a continuous foreshore for the free drift of the shingle, and A. J. Burrows notes that for this to be the case the Brede and Tillingham must have had their outlets far to the east. Drew agrees with the above authorities concerning the run of the coast from Hythe to Lydd, but from that place, instead of continuing his coast straight to Fairlight, he suggests[1] three positions, passing through *F*, *L* and *N* respectively, and then turning westwards to the neighbourhood of *C*. It seems probable that the shoreline at different times followed all three of these positions, that through *F* being the earliest, and those through *L* and *N* being successively later.

HISTORICAL EVIDENCE

It has been generally held in the past that the Rother reached the sea at Hythe in Roman times (M. Burrows, 1888; Drew, 1875; Elliot, 1847; Holloway, 1849; Lewin, 1862; Topley, 1885), and that later its course was diverted to New Romney. Dowker (1897–8)

---

[1] One-Inch Geol. Surv. Maps, Sheet 4.

and Roach Smith (1852) deny this suggestion, and the former considers that the error is due to the confusing of Lemanis, which was situated near Stuttfall Castle, west of Hythe, with Portus Lemanis, which was on the Rother but probably 10 miles distant, near New Romney. The physical history[1] of Romney Marsh is against the existence of a Hythe outlet for the Rother, because the marsh has always drained southwards towards the Rhee Wall, and evidence is completely lacking of a former channel at the foot of the hills. It seems therefore that the early inlet, which certainly existed near Hythe must not be confused with the mouth of the Rother, which at least since Roman times has been at New Romney or farther west.

According to Lewin (1862, p. 221) the Romney Marsh was drained by the building of the Rhee Wall, which included a channel 80 to 100 ft wide, running from Appledore to Romney. More recently, however, Dowker (1897–8, p. 215) has shown that the land in all the settlements of East Kent was higher in the Roman era than today. This must have greatly facilitated the draining of the Marsh, and C. J. Gilbert (1930, p. 97), on good authority, states that without this uplift the Rhee Wall cut would not have accomplished its purpose.

In the course of the eighth century a very important change began to take place. Holloway (1849, p. 49) states that by A.D. 774 Lydd existed on an island at the mouth of the Rother, but that by 791 the sea seemed to be retreating fast. Montague Burrows's (1888, p. 14) views on these changes at Lydd are most instructive. He refers to Eadbriht's Charter of 741 and Offa's of 774 making grants of marsh land near Lydd to the Church of Christ in Canterbury, and concludes that 'these charters show that a change had taken place which we can now identify with the growth of Dungeness towards the sea. . . . Lydd however was still surrounded by the sea on the north and east, and in 893 the piratical Danes made their way past it up to Appledore with a fleet of 250 vessels.'

An important result of this extension was the holding-up of shingle which previously fed the foreshore on the present site of Dymchurch Wall. Elliot (1847, p. 470) and Redman (1952–3, p. 196) both account thus for the decrease of shingle along this part of the coast, a decrease which eventually necessitated the organised system of protection which culminated in the construction of the present Dymchurch Wall by Rennie in 1803–4.

---

[1] The writer follows Lewin in applying this name only to that part of the marsh to the north and east of the Rhee Wall.

W. V. Lewis

In the thirteenth century the third and last great great change in the evolution of Dungeness took place, i.e. the destruction of Old Winchelsey and Promehill, and the diverting of the Rother from New Romney to the neighbourhood of its present outlet at Rye (Fig. 4.1).

According to Cooper (1850, p. 1) the site of Old Winchelsey was a low flat island, 6 miles north-east of Fairlight Cliff, 3 miles south-east by east of modern Winchelsea, 2 miles south-south-east

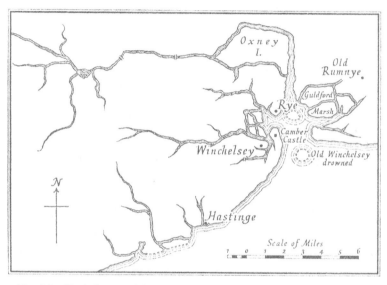

Fig. 4.3   Norden's map of the East Sussex coast, showing site of Old Winchelsey

of Rye, and 7 miles south-west of Old Romney, and he states that Jeake described it as being washed by the 'British Ocean' on the south and east and the mouth of the Rother on the north. Old maps, such as Norden's, of the coast of East Sussex (Fig. 4.3) also show it roughly in this position, but the site is not accurately known.

The most enlightening reference to the early conditions at this part of the coast is due to Lewin (1862, p. cxix): 'It is highly probable that both Old Winchelsey and Promehill stood, as Lydd does, on an ancient shingle spit, and not higher above high-water mark than three to four feet. ... The sea probably, at some earlier period of which we have no record, severed the spit somewhere between Old Winchelsey and Promehill, on the Camber Galles. ... There cannot be much doubt that at one period the shingle spit formed a communication across the bay, part of the parish of Winchelsea being still on the east side of Rye Bay and extending to a point on the coast where a man can see Beachy Head.' This severing of the spit to which he refers suggests that this section of the coast was being driven back by the sea from a very early date.

The culmination of this movement, however, was the complete destruction of Old Winchelsey and Promehill by the waves and the last great diversion of the course of the Rother. This catastrophe has been described by all the earlier chroniclers (Chronicles; Holloway, 1849; Lewin, 1862; Somner, 1693) and is summed up by the following quotation from Montague Burrows (1888, p. 114): 'In the fearful crash and inundation of 1287 the old half-ruined town was entirely swept away; the course of the river Rother was changed and with it the whole face of the Romney coast. For many centuries no one has been able to point to any particular spot with certanty and say "Here stood Winchelsey".'

The fate of New Winchelsea, built by Edward I to house the people driven from the old town, indicates yet another phase in the changes of the coastline. Concerning Winchelsea, Montague Burrows (1888, pp. 115, 221) writes that Edward had 'evidently designed to make his new town the headquarters of the Cinque Ports, little foreseeing that it was about to be ruined by the stealthy retreat of the sea as the old one by its tumultuous advance'. Further light is thrown on this retreat of the sea by the building of Camber Castle to defend the entrance to Winchelsea harbour: 'Henry VIII made up his mind from personal inspection that the place [Winchelsea] was no longer of any use for his continental wars, and as a substitute, built Camber Castle at the head of a small estuary or camber, to which the old harbour had shrunk. He was deceived in his turn as completely as his great predecessors; for Camber soon took, though gradually, the way of the former harbour, and in the course of the century left his castle high and dry two miles from the sea. There it still stands, a ruinous monument of a third failure on this fated spot.' Norden's map (Fig. 4.3) represents the state of the coast soon after the building of Camber Castle.

The recent changes at Hythe, apart from the decrease of shingle on the site of the Dymchurch Wall, are connected with the decay of Hythe haven. It is a debatable point as to whether or not the Portus Lemanis of Roman times (M. Burrows, 1888; Dowker, 1897–8; Holloway, 1849) was near the present Lympne, but that an outlet existed there is agreed, though it gradually moved eastwards owing to the extension of the shingle ridges to West Hythe and Hythe before it was finally closed. Holloway (1849, p.137) records that in the reign of Henry VIII the haven of West Hythe seems to have been destroyed by the sands and beach cast up on the shore, 'for Leland describes it as being, at that time, only a small channel or gut left, which ran within the shore for nearly a mile eastward from Hythe towards Folkestone, where small vessels could come up with safety'. By Elizabeth's reign, however, all this was lost.

We have therefore a valuable historical account of the changes which embody the growth of a considerable portion of Dungeness. In the following pages the writer will suggest reasons for these changes which are known to have taken place and incorporate them in a suggested explanation of the evolution of the foreland.

The last considerable movement of the land in the neighbourhood of Dungeness was the subsidence known as the Neolithic Depression. The resulting shoreline probably followed closely the old cliffline shown in Fig. 4.1, and there was then no shingle foreland or tract of marshland. The rivers Rother, Tillingham and Brede then entered this great shallow bay through wide tidal estuaries. The Battle Ridge, composed of relatively resistant Ashdown Sands, in jutting out to the south-east as Fairlight Head, brought about the change in direction of the old shoreline from east-north-east, near the present site of Hastings, to north-north-east, along the western shore of the Romney embayment.

In estimating the position along this coast at which deposition would first take place, the writer inspected this bend in the shoreline at Fairlight. An examination of the Camber Castle ridges which have developed to the lee of this headland not only suggested reasons for the beginning of the deposition of Dungeness, but also indicated a systematic change in the direction of the ridges which provides the key to the solution of its later development.

The prevalent south-west waves drift large quantities of shingle, mostly flints derived from the chalk cliffs of Sussex, eastwards to Fairlight. Just at this headland, where the old cliffline changes its direction from east-north-east to north-north-east, a shingle ridge has been built out from the shore to the north-east. That is, the ridge nearly continues the direction of the coast to the windward, and also makes a smaller angle with the dominant waves, which are southerly at this point, than does the old cliffline behind it (Lewis, 1931, p. 134).

The disposition of the ridges is shown in Fig. 4.4, taken from Sheet 4 of the One-Inch Geological Survey Maps; only those areas consisting of bare shingle are mapped, but on the ground the ridges can often be traced across the intervening gaps. For nearly two miles from Cliff End the shoreline consists of a single embankment of shingle, which, in spite of heavy groyning, is rapidly being thrown

Fig. 4.4 Disposition of shingle ridges at Camber Castle

back by the waves on to the marshes behind. The coast road from Winchelsea to Cliff End is now impassable owing to the encroachment of the shingle. Along this section the ridge has been thrown back beyond the ends of any lateral ridges that might have existed. East of *A*, however, many of these old ridges are visible: an early one runs from *A* through *B* to *C*. By the addition of ridges overlapping the end of the spit this shoreline was extended northwards and eastwards until the stage *ABDM* was reached. Following this, a completely new ridge started to grow northeastwards from the neighbourhood of *A* to *E*, leaving a narrow depression between it and the older shoreline *BD*. By a similar addition of overlapping ridges a continuous shingle foreland was built forward to *AF*, following which a second set of independent ridges started from *A* and eventually reached *K*, enclosing a second well-defined depression known as Nook Creek. Continuous addition of shingle since the time of the Geological Survey has resulted in the present shoreline following the line *AL*, the advance of *L* being accelerated by the construction of high groynes which have to be extended continually in order to preserve the channel of the Rother for navigation.

The mode of extension of these ridges favours the suggestions put forward by the writer in his previous paper (Lewis, 1931, p. 137), but the action which is the most important in the present discussion is the tendency for each new shoreline to pivot around a point somewhere to the south-west of *A* and face more to the south, the direction from which the dominant waves approach. The writer suggests that such a swinging round of the ridges so as to face a more southerly direction is the basic factor in the evolution of Dungeness.

At the close of the Neolithic Depression there seems no reason to believe that conditions of wave incidence and shingle supply should be materially different from those obtaining today north-east of Fairlight Head, with the exception of somewhat increased exposure from the east owing to the absence of the projecting foreland of Dungeness. It therefore seems logical to conclude that the earliest ridges, which were built and later destroyed, would have closely resembled those just described, that is, would have extended north-eastwards from a Fairlight Head which stood about a mile to the seaward of its present site.[1] They are represented by *AB* in Fig. 4.1, and ended somewhere south of the present mouth of the Rother. In swinging round to face more nearly the dominant southerly waves, the movement referred to above, the successive shorelines occupied positions represented by *AC* and later *ADE*. The line *DE* represents a series of ridges which are actually present on the foreland. They therefore comprise the oldest ridges which are still preserved. The next shoreline, *AFG*, continues through Lydd and New Romney right to Hythe, and is the earliest shoreline referred to by the authorities quoted above. In this case it is necessary to account for the one particular shoreline extending longitudinally nearly a dozen miles beyond the ends of the previous ones.

The explanation, which was hinted at by Dowker (1897–8, p. 221) and definitely put forward by C. J. Gilbert (1930, p. 93),

---

[1] De Lamblardie's estimate of a recession of 1 foot per annum.

seems to lie in the change in the relative level of the land and sea in Submerged Forest times. Mr Gilbert informs me that the uplift of the land for the deposition of the Forest Bed must have been at least 25 ft, and that a later uplift for the Roman occupation of the Marsh was very much less, but its exact amount has not yet been determined.

If, then, we picture the changes resulting from this first uplift of the land, the reasons for the extension of the old shoreline towards Hythe are evident. The land must have risen above the present level before the ridges *DE* (Fig. 4.1) were built, for the writer observed that these are about 8 ft below the level of those comprising the present foreshore. When the full 25-ft elevation took place, much of the sandy areas which previously formed the offshore bottom was lifted above sea-level, and where there was once deep water was now shallow. There had probably been a considerable deposition of sand and fine material on the site of Romney Marsh owing to the protection from the south and south-west waves afforded by the shorelines leading up to the stage *ADE*.

Over this shallow area, then, a shoreline extended towards Hythe. How far it had progressed before the land began to return to its original level has not yet been determined, but Elliot (1847, p. 467) and Appach (1868, p. 21) refer to the presence of recurves into and under the marsh on the site of the Dymchurch Wall, which must have been formed when the sea-level was considerably lower than at present. On approaching Hythe, however, there is an extensive shingle foreland composed of the successive recurved ends of this old shoreline, which are within a foot or so of the level of the ridges forming the present foreshore.

This shoreline probably extended very slowly eastwards from a point some distance north-east of Dymchurch, because the opening which history records as being at Hythe in Roman times was not closed by the eastward growth of the shingle ridges until Elizabeth's reign. This agrees with the evidence, noted by the writer, of the change in level of the old ridges by comparison with those forming the present foreshore. It has already been remarked that the recurved ridges near *G* were built when the sea was at its present level, but according to Oldham this did not happen until the thirteenth century. Therefore a great length of time had elapsed before the *G* shoreline, begun in Submerged Forest times, eventually reached Hythe.

It is now necessary to consider the fate of the part of this shoreline which was built when the land was at its highest level, for, on submergence, one might at first expect it to be overwhelmed. The subsidence of the land, however, was apparently so gradual that the ridges were built up by the waves as quickly as the sea-level rose, at the same time being driven landwards a distance which cannot yet be determined. When the movement ceased the shoreline probably followed the line through *F*, Lydd and New Romney. An examination of the section still preserved near New Romney suggests that it consisted of parallel ridges of mixed sand and gravel[1] which formed only a narrow barrier between the sea and a large lagoon, the main outlet of which was between Hythe and New Romney.

The succeeding stage was perhaps the most important one in the evolution of the foreland; it led to the formation of the bend (*K*, Fig. 4.1) which eventually developed into the ness form of the present day. As the sea returned to a higher level and drove back the Forest ridges, it seems to have broken through this shoreline into the lagoon behind at two points. The first, near the present site of New Romney, then became the outlet of the Rother, and through the second, to the north-east of Fairlight Head, the Tillingham and Brede thenceforward made their way to the sea. Such gaps must have been formed at some early date in order to have allowed the rivers to reach the sea at these points as they did by Roman times.

The shingle to the east of the Fairlight estuary would still be drifted to the north-east, but further supplies, in having to cross the river channel, would be reduced, so that the shoreline, starved of material in this manner, would suffer erosion. The ridges of a shoreline suffering erosion are being continually re-shuffled by the storm waves, with the result that these latter have full opportunity to turn the ridges into a direction more nearly parallel with their own fronts. Thus the shoreline, with the western end no longer attached to Fairlight, would tend to swing round to face more nearly the dominant southerly to southwesterly waves.[2] This would most easily be accomplished by its assuming the form *HKL*, similar to the earliest shoreline suggested by Drew.

As the southern shore was driven back in this manner the sharpness of the bend at *K* increased, and this had important consequences on the subsequent development. The sharpening of this bend led to

---

[1] Good exposures are visible from the main road from Lydd to New Romney.
[2] The direction of the largest waves reaching this coast has been ascertained by inquiries from the local fishermen.

conditions which resemble those obtaining at the Ness today, so that a study of the latter throws light on the problem.

When the bend was slight it had little effect upon the rate of drift, but as it became sharper the drift along the leeward shore was lessened. For the prevalent south-west waves which cause this drift are so weakened in swinging round the bend that in spite of their greater obliquity they are unable to drift material northwards from the Ness as rapidly as they bring it along the southern shore (Drew, 1875, p. 17). This results in large supplies of shingle accumulating immediately round the point, which in turn is built into ridges overlapping the point by the south-west waves, thus causing the Ness to advance seawards. This explanation, given in the case of Hurst Castle Spit (Lewis, 1931, p. 137), was deduced from an examination of the speed of drift on either side of the point and the plan of the ridges which had been added recently. Since that time the writer has been fortunate in obtaining at Dungeness direct evidence in favour of this mode of spit extension.

When the writer visited the Ness in September 1929, the curvature at the point was similar to that of XY (Fig. 4.5). By June 1931 the outline was similar to that of $ADBZ$, and it was realised that the key to the eastward extension of the Ness lay in a true interpretation of the processes which had formed this salient at B.[1] These processes were clearly indicated by an examination of the ridges comprising this part of the foreland.

The ridge XY showed no sign of a salient, but the next ridge ABC presented one in a more highly developed form than the shore-line at the time of observation, so that it was the building of this ridge which had occasioned the extension at B. The ridge had been driven back nearest the one behind and was of greatest height along the section DE, suggesting that it had been built by storm waves which approaches dead onshore to this section. Beyond E it turned more to the north-east, and was slightly reduced in height. From E to B it extended farther and farther seaward from XY, eventually leaving a deep trench between the two. Finally the ridge curved back, still more reduced in height and bulk, and joined XY at C.

[1] The writer is fully aware that the salient described is the result of a particular set of storm waves and that such waves from another direction, e.g. the east, might build a temporary salient pointing southwards. The one under consideration might be partially or even wholly removed at some future date, but the fact that it has lasted for two years and that similar salients are preserved in the ridges just behind the foreshore suggest strongly that they play the major part in the eastward extension of the Ness.

This addition seems to have taken place in the following manner. Large storm waves approaching dead onshore to the section *DE* removed large quantities of material down the beach and threw some over the top to increase the height of the ridge. As the shore bent round beyond *E*, the waves found an abundant supply of material in the lesser ridges comprising the upper foreshore, and presumably received much from the erosion along *DE*. This they threw back into a ridge *EB*, continuing in the direction of the waves

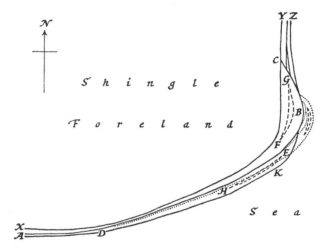

Fig. 4.5 Extension of Dungeness, 1929-31

for some distance beyond the bend in the previous ridge before being driven back to rejoin the foreshore, enclosing the trench *FG*. The presence of this trench, into which water percolated at high tide, shows clearly that here the amount of shingle was sufficient to form a ridge massive enough to resist being driven right back on to the previous ridge by the southerly breakers. To extend beyond *B*, however, it would reach farther down the foreshore to deeper water where the increasing power of the swash of the storm waves together with a reduced supply of shingle would cause the ridge to be driven back on to the earlier one at *C*, a movement made easier by the supply of shingle on the foreshore becoming smaller as the distance north of the Ness increased. This swinging round of the breakers to drive the ridge back to *C* was made possible by the great depth of water immediately offshore which allowed the seaward ends of the breakers to continue their forward motion whilst the landward ends were retarded on encountering the shore. A similar action due to waves from a slightly more southerly direction has caused another ridge to be built along *DK*, again leaving a deep trench *HE* between it and the ridge *DB*. The addition of these two ridges has extended the Ness from the line *XY* to *AKBZ*.

Between June and December 1931 the Ness had extended eastwards a further 20 yds by the addition of other ridges overlapping the point indicated by dotted lines in the figure. These last ridges, however, were distinctly lower than the earlier ones, and represent only a temporary addition owing to an unusual frequency of south- westerly winds in the preceding months. They will be removed by powerful easterly winds and the shingle will be redistributed along the foreshore to the westward. Therefore these instances afford clear examples of the extension of a spit or embankment of coarse material into deep water, in a direction largely determined by that of the storm waves responsible for the addition. These, then, seem to be the conditions which led to the eastward extension of the Ness as it sharpened in the stages following *HKL* (Fig. 4.1). It now remains to consider what happened to the shingle which travelled round the Ness to the northward.

As in the case of Hurst, the direction of the lateral ridges formed from this shingle was determined by that of the dominant storm waves for this section of the coast. The southwesterly and southerly waves are very much reduced in size after swinging round to reach this lee shore, and can therefore be neglected. To the south-east lies only a limited stretch of water, but to the east-north-east between the South Foreland and Cap Gris Nez open water extends right into the North Sea, and so it is from this direction alone that large storm waves can reach the eastern shores of Dungeness.[1] The writer observed the front ridge of the foreshore *YC* (Fig. 4.5) actually being built in this manner by easterly waves.

[1] In reply to inquiries as to the largest waves developed in the Channel, Mr James T. Blake, Divisional Marine Manager to the Southern Railway, Marine Department, Folkestone Harbour, and Mr J. W. Oiller, a resident of Dungeness, kindly gave me the following information. The largest waves in the Channel, after excluding those from the south-west, are from northeasterly and easterly gales.

Mr J. P. Taremy, Station Officer, H.M. Coastguard, Lade, Lydd, Kent, replied that the largest waves reach the eastern shore of Dungeness from the north-east to east-north-east, and that the next largest come from the south to south-south-east; thus confirming the writer's conclusions that the direction of approach of the largest waves is controlled by the available stretch of open water.

Thus similar circumstances would suggest that the shingle carried northwards from the Ness at K (Fig. 4.1) would be built into ridges tending to face the dominant east-north-easterly waves. Therefore as the southern shore extended eastwards the lateral ridges no longer remained parallel with the earlier ones, but turned more and more to the north-east, a process which resulted in their northern ends being driven back to join the old shoreline at Lydd. This led to still further sharpening of the Ness and to a corresponding strengthening of the conditions favouring its eastward extension.

In time a stage represented by a line passing through M, N and Lydd was reached, the shingle of the ancient ridges AB, AC, etc., being drifted eastwards and deposited as the lateral ridges north of N. By this time the lateral ridges faced east-north-east, and so in the later stages they tended to maintain this direction. The Ness had now reached a form very similar to that shown at the present time. The reason for this form is, then, the sudden change in exposure at the point; large waves approach either from the south or from the east, so that the two shores face these two directions. The sharpness of the point is due to the proximity of the French coast preventing the occurrence of large waves from a south-easterly direction, which would tend to flatten the point. A similar feature is present at Hurst, and the writer has noted several smaller forelands on both sides of the Bristol Channel, near Barry on the Glamorganshire coast, and Minehead on the Somersetshire coast, where the points always face the direction in which lies the smallest stretch of water. It seems therefore that the form of these cuspate forelands can be adequately explained in terms of this sudden change in exposure, allowing large waves to approach at an oblique angle with the general trend of the coastline from two different directions, but never directly onshore.[1] These conditions would be most frequent in long narrow channels, and although G. K. Gilbert (1885), Gulliver (1899), Wilson (1904), Tarr (1898) and Woodman (1899), amongst others, were familiar with this feature of the distribution of these forelands, they did not grasp its significance, and were therefore unable to explain them satisfactorily.

All the lateral ridges west of N are from 5 to 8 ft below the level of the modern shore ridges, but at the stage MN they rise quite suddenly to this level. R. D. Oldham suggests that the sea returned

[1] In this generalisation the writer excludes Orfordness from the category of typical cuspate forelands as it has developed at a point where the coastline changes its direction.

to its present level in the thirteenth century, making this the date at which the Ness followed the outline *MN*.

The manner in which the sea deserted Lydd at the beginning of the eighth century has already been described, and if this was associated with the extension of the Ness eastwards from *N*, as suggested by Montague Burrows, this is hardly in agreement with the date given by Oldham. As more evidence becomes available, however, this discrepancy between the age of the ridges at *N* deduced from the views of the above two authorities will probably be explained.

After this time the eastward extension was hastened by the bend having reached its sharpest at *N* and was too rapid for the lateral ridges to continue far to the north before having new ones built in front of them. As the Ness continued eastwards into deeper water, its extension seems to have become slower, for the lateral ridges have extended farther and farther north to reach their maximum development in the present shoreline. From the ninth to the thirteenth centuries much land south of the Rhee Wall was inned by the successive Archbishops of Canterbury (M. Burrows, 1888, p. 15) and this must have caused the Rother to follow a more southerly course to New Romney (Fig. 4.1), if it had not done so from an earlier date.

Old Winchelsey was probably built on a shoreline similar to *BQR*, its sudden rise to power together with Rye presumably being due to the changing conditions offering a suitable site. These changing conditions might well have been the breaking through of the spit, leaving Winchelsey exposed to the sea on the south and east and with the Rother (apparently the combined outlet of the Tillingham and Brede) on the north as described by Jeake (Cooper, 1850). This could also give Rye a good harbour to the south of the town. Even if such were not the cause of Winchelsey's sudden prosperity, a break in the spit certainly took place in its early history, leaving a large portion of its parish on the mainland near Bromhill and transforming its site into an island.

The later history seems most instructive concerning the northward movement of the old shoreline. Eventually the advance of the sea, after threatening Winchelsey for the greater part of the thirteenth century, caused its complete ruin and also broke through the rampart of shingle into the marshes to the north, opening a new mouth for the Rother. This spot would naturally be a weak point in the shoreline, owing to the absence of old lateral ridges which are present farther east, whilst shingle would be readily removed by the south-west waves, but only slowly supplied by crossing the estuary of the Tillingham and Brede.

This final change was first attributed to a period of great storminess in the eleventh to thirteenth centuries (Chronicles; Holloway, 1849; Lewin, 1862; Redman, 1852–3), which wrought much havoc on several parts of the English and Dutch coasts. A more feasible explanation, suggested by R. D. Oldham (Steers, 1927) and elaborated by C. J. Gilbert (1930, p. 99), is that the final stage of the depression of the land following the uprise for the period of the Roman occupation took place just prior to the thirteenth century. There is little doubt that this depression could have brought to a head the destruction of Old Winchelsey, but it seems that it only accelerated the changes which must have been the inevitable result of the swinging round of the southern shore. The augmented outflow of the Rother, Tillingham and Brede after the thirteenth century held up large quantities of shingle to the windward. This was built into the Camber Castle ridges already described, an addition which brings the foreland to its present form.

The above explanation seems to agree both with the historical facts and the observations made on the foreshore and studied in the light of the theories on spit-building outlined in the author's previous paper to this Society. He is fully aware that certain modifications will probably have to be made as more evidence becomes available, but the ease of interpretation of the various phases in the development of this important coastal feature does seem to indicate that this suggested explanation is in the main correct.

REFERENCES

APPACH, F. H. (1868) *Caius Julius Caesar's British Expedition and the Subsequent Formation of Romney Marsh.*

BURROWS, A. J. (1884–5) 'Romney Marsh, past and present', *Surv. Inst. Trans.*, XVII 338.

BURROWS, M. (1888) *Cinque Ports.*

CHRONICLES. Leland's *Itinerary* (1539); Holinshed's *Chronicles* (1577); Camden's *Great Britain* (1586).

COOPER, W. D. (1850) *History of Winchelsea.*

DOWKER, G. (1897–8) 'On Romney Marsh', *Proc. Geol. Assoc.*, XV 215.

DREW, F. (1875) 'Geology of Folkstone and Rye', *Mem. Geol. Surv. Gt. Brit.*

ELLIOT, J. (1847) 'Account of Dymchurch Wall', *Mins. Proc. Inst. Civ. Eng.*, VI 466.

GILBERT, C. J. (1930) 'Land oscillations during the closing stages of the Neolithic Depression', *2nd Rep. Comm. on Pliocene and Pleistocene Terraces, Union Géographique Internat.*

GILBERT, G. K. (1885) 'Topographic features of lake shores', *5th Ann. Rep. U.S. Geol. Surv.*

GULLIVER, F. P. (1897) 'Dungeness foreland', *Geogr. J.*, IX 536.

—— (1899) 'Shoreline Topography', *Proc. Amer. Acad. Arts & Sci.*, XXXIV 214.

HOLLOWAY, W. (1849) *History of Romney Marsh.*

LAMBLARDIE, DE (1789) *Mémoire sur les côtes de la Haute Normandie.*

LEWIN, T. (1862) *The Invasion of England by Julius Caesar.*

LEWIS, W. V. (1931) 'The effect of wave incidence on the configuration of a shingle beach', *Geogr. J.*, LXXVIII 141.

REDMAN, J. B. (1852–3) 'On the alluvial formations and local changes of the south coast of England', *Mins. Proc. Inst. Civ. Eng.*, XI 186.

ROYAL COMMISSION ON COAST EROSION (1911) *Final Volume.*

SMITH, C. R. (1852) *Report on Lymne* (published privately).

SOMNER, W. (1693) *Roman Ports and Forts in Kent.*

STEERS, J. A. (1927) 'The East Anglian coast', *Geogr. J.*, LXIX (see R. D. Oldham in discussion, p. 46).

TARR, R. S. (1898) 'Wave-formed cuspate forelands', *Amer. Geol.*, XII 1.

TOPLEY, W. (1885) 'Geology of the Weald', *Mem. Geol. Surv. Gt. Brit.*

WARD, E. M. (1922) *English Coastal Evolution.*

WHEELER, W. H. (1902) *The Sea Coast.*

WHITE, H. J. O. (1928) 'Geology of the country near Hastings and Dungeness', *Mem. Geol. Surv. England and Wales.*

WILSON, A. W. G. (1904) 'Cuspate forelands in the Bay of Quinte', *J. Geol.*, XII 132.

WOODMAN, J. E. (1899) 'Shore development in the Bras d'Or Lakes', *Amer. Geol.*, XXIV 329.

5 The Coast of Louisiana

R. J. RUSSELL

LOUISIANA is located centrally on the east–west coast north of the Gulf of Mexico. From the standpoint of structural geology it lies on the northern flank of the Gulf Coast geosyncline. This major structure started to form during the Lower Cretaceous, and it has remained a site of active deposition ever since. During the Neogene and Quaternary southern Louisiana has experienced tilting towards the geosynclinal trough. Actual uplift in the north has accompanied actual subsidence in the south.

The amount of tilting during comparatively recent time may be appreciated from some of the following data. In oil wells of coastal Louisiana, upper and middle Miocene are ordinarily the oldest rocks encountered, even at depths of more than − 4,000 m. These beds outcrop in central Louisiana at such elevations as 100 m., 300 km. north of the coast. An oil well approximately 45 km. south of the shoreline, on the shallow continental shelf, 'bottomed' in middle Miocene at − 3,930 m. Another well about an equal distance inland reached a depth of − 4,680 m. without penetrating beds as old as lower Miocene. The base of the Pleistocene is encountered at about − 750 m. by wells on the continental shelf, and the base of the Holocene at about − 183 m.

Louisiana geologists regard the Recent, or Holocene, as the period during which the last major rise in sea-level occurred. This definition gives stratigraphic validity to the Recent as a period, or 'stage', of Quaternary time.

The base of the Recent is distinctly marked in cores from wells in coastal Louisiana (Fig. 5.1). All of the Recent section is composed of sediments that have been in an environment favouring reduction ever since deposition. Peat is relatively abundant. Clays and silts predominate in the upper part of the section, though there are many irregular lenses of sand or shells. Towards the base of the Recent are coarser sediments, with abundant sand and, across the shelf, minor amounts of gravel. Capping the Pleistocene is a conspicuous zone of material oxidised during the last major low-stand of Pleistocene seas. The black or dark-bluish Recent stands in sharp contrast to the yellowish or reddish capping of the Pleistocene. Calcareous and iron–manganese nodules are abundant in the oxidised zone.

The base of the Recent has been cored at hundreds of places. Many samples have been preserved of the Recent–Pleistocene contact across both the marshes of coastal Louisiana and the continental shelf. The break is known at so many places that Dr Harold N. Fisk, of Louisiana State University and the Mississippi

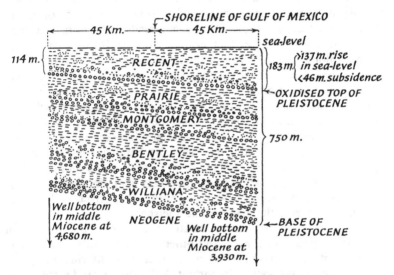

Fig. 5.1 Quaternary section in coastal Louisiana

River Commission, has been able to map in detail the pre-Recent drainage channels leading to the Gulf of Mexico. This supplements the published map which shows the depth of the Recent–Pleistocene contact under the entire Lower Mississippi valley, between Cairo, Illinois, and a line extending from Franklin to Donaldsonville. Louisiana (Fisk, 1944, Plate 3). Though several hundred penetrations of the pre-Recent surface have been recorded since the map was completed, in about 1942, and the contour interval is 25 ft above sea-level and 50 ft below, with the lowest contour at −350 ft, no significant change would be made in the position of any contour were the map being reissued today. The topography upon which the Recent was deposited is thus known in detail. A statement that the volume of Recent alluvium in the Lower Mississippi Valley,

between Cairo and Franklin, is 4,160 cu. km. is probably subject
to an error of less than 10 per cent.

A projection of the gradient of the pre-Recent Mississippi
valley to the outer edge of the continental shelf, where it reached
the Gulf of Mexico during the pre-Recent, low-stand sea-level,
indicates a maximum depth of about − 137 m. It thus appears that
the Recent–Pleistocene contact at the depth of − 183 m. in a well
45 km. south of the shoreline represents something of the order
of a 137-m. rise in sea-level and about 46 m. of actual subsidence.

The southward tilting of coastal Louisiana is clearly demonstrated
in the Pleistocene section. Four widespread major terraces extend
down the Mississippi and other rivers of the Gulf Coast region.
Those along the Mississippi maintain rather constant vertical
separations southwards to about the northern boundary of Louisiana.
In crossing that state the upper terrace is warped downwards, so
that it crosses the other three. The Williana formation, which consists of the oldest and highest terrace deposits inland, becomes the
lowest formation in the deltaic stratigraphic section of southern
Louisiana and the continental shelf. Its basal gravels overlie the
Neogene at a depth of − 750 m., 45 km. south of the shoreline. In a
similar manner the Bentley, Montgomery and Prairie formations
complete the Pleistocene section, both along river valleys and
southwards across the marshes and the shelf.

Each of the Pleistocene formations resembles the Recent in
consisting basally of gravels and other coarse deposits which grade
upward into silts, clays and other fine materials. 'Marine' intercala-
tions are comparatively unimportant. Valley and inland sections are
typically continental. Coastal and shelf sections are palustral and
deltaic, rather than typically marine.

Though Tertiary tilting has much less influence than Quaternary
subsidence on existing coastal forms, it is interesting to note that the
oil wells of coastal Louisiana remain in the Miocene at depths which
are actually more than 1,000 m. below the deepest part of the floor
of the Gulf of Mexico. It is also interesting that the Neogene section
is almost entirely deltaic, rather than typically marine.

The morphological details of the shoreline of Louisiana are determined by three principal factors: tilting, which results in coastal
subsidence; alluviation; and wave erosion. Though waves encounter
only unconsolidated Recent sediments along the entire coast, four
distinctly different kinds of shoreline have been developed.

The Plaquemines (Parish) shoreline is one dominated by alluviation (Fig. 5.2). It is the shore of the active Lower Mississippi river delta.

The origin of the 'bird-foot' delta pattern has long been a matter of speculation. Theories that the form may be related to tideless sea, to feeble littoral currents or to certain directions of prevailing

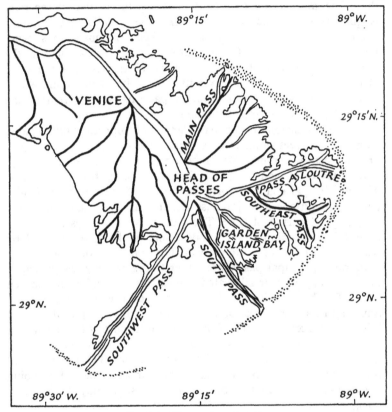

Fig. 5.2 Plaquemines shoreline

or dominant storm winds are unsound. Several earlier, but strictly 'modern', Mississippi deltas exist in relict form, but none has the bird-foot pattern. This peculiarity appears to be related to the more or less accidental fact that the river is now confined to a single channel, without either tributaries or distributaries, for its lower 400 km., to Head of Passes. In that region it suddenly branches into several distributary outlet channels. The longest of these, Southwest Pass, has a length of about 33 km.

The important passes of today are not those of A.D. 1800. That they date from about 1892 is due chiefly to the activities of engineers who are interested in maintaining them for purposes of navigation. For a distance of about 160 km. below New Orleans the Mississippi channel is cut in 'blue' clay deposits of its older deltas. In these fine-grained and tenaceous materials the course of the river is quite straight. The channel is 'fixed', and shifting or channel abandonment is not readily accomplished. A conspicuous bend, English Turn, near New Orleans, owes its origin to the fact that the river crosses the Lake Borgne fault zone. Near Venice, in the Head of Passes region, the river leaves its older delta deposits and is free to create its own, new distributary patterns. It is here that the typical Plaquemines type of shoreline finds free development.

Each new distributary, or 'pass', grows out into the Gulf rapidly, with natural levees along its sides. Principal distributaries are straight, or only gently curving. Minor channels below crevasses through natural levees of major channels form interlacing, braided patterns. A few principal channels form the talons of the bird-foot. Under natural conditions a major pass remains an important outlet for only a short period of time. The passes used by ships in 1700 or even 1800 are unimportant today. Some of the channels used by sea-going ships in 1850 are alluviated and have become obscure.

Between principal passes are either triangular bays, with depths of 2 or 3 m., or their alluviated equivalents. The filling of Garden Island Bay by silty deposits below Pass à Loutre crevasse, between South and Southeast passes, has been practically completed, though initiated as recently as 1890. Charts are misleading at that particular place. Water appears to cover about half of the original area of the bay, but places covered to a depth of 3 m. in 1890 are too shoal for shallow-draught skiffs today. Thus arises a partial webbing between some of the talons of the bird-foot, suggesting the foot of a duck.

Charts are also deficient in showing the bars and beaches which are submerged at high tide and which are nearly continuous around the periphery of the Plaquemines delta. This deposit of find sand, silt, shell and whatever coarse materials are available to wave currents is tangential to the distal ends of passes, so that it practically forms an arc of a circle. Where it crosses channels, dredging is necessary if navigation is to be maintained. On South Pass, jetties have solved the problem of bar accumulation quite well. They have been less successful on Southwest Pass. These are the only passes available to ships drawing more than 1.5 m. Main Pass commonly presents difficulties to fishing boats drawing 1.3 m. or less. Other passes, such as à Loutre and Southeast, are at times of river flood insufficiently deep for skiffs drawing but 0.2 m. They deepen slightly after floods subside because finer sediments are winnowed from their bars. The bottoms then become harder, sandier and in some ways more dangerous for smallboat use.

The highest lands in the Lower Delta are strips along the crests of natural levees whose elevations are markers of flood crests. At Head of Passes the elevation of land attains a maximum of about 1.8 m. Heights of the comparatively dry natural levee crests are magnified by willows and other tall trees which tower above the grasses of the surrounding marshes. Where the dry silts of natural levees are sufficiently broad, they are protected by artificial levees and form sites for agricultural use. Roads and artificial levees extend south to Venice. Head of Passes may be reached only by boat or seaplane.

Natural-levee fine sands and silts grade into heavy, organic clays in the 'backswamp', or low, grassy marshes between distributary channels. Fine sand, silt and finer materials travel gulfwards along principal channels. Below crevasses, fine sand and silt are commonly deposited as widespread sheets or as irregular lenses which become intercalated in the backswamp clays. Other beds of coarser sediments accumulate on the floors of shallow lakes within the marshes. On the whole, delta sedimentation is complex. On large deltas there is no particular merit in the concept of top-set, fore-set or bottom-set beds. In general, as on many flood plains, the most distinct contrasts in sediments occur transversely with regard to channel directions.

In summary, the Plaquemines type of shoreline is one dominated by alluviation. Its irregularities are caused by river deposition, but the smoothing influence of wave action is evidenced by the presence of a low, peripheral beach which is growing as an arc between the ends of distributary channels.

The St Bernard (Parish) type of shoreline (Fig. 5.3), which is developed in the broad area north of the Plaquemines region and generally east of New Orleans, exhibits early effects of the abandonment of active alluviation. Possibly as late as the tenth century the main course of the Mississippi extended eastwards from New Orleans into the St Bernard region. A comparatively dense population of Indians inhabited the natural levees of the St Bernard delta.

Artefacts from their mounds and middens place the latest Indian culture well toward historic time (locally, since A.D. 1600), but no finds of European 'trade articles' have been found. There is an absence of fresh water today in the distal remnants of the St Bernard delta, where bases of Indian mounds lie submerged 3 m. and more below present sea-level.

Against the idea that sea level has risen appreciably since the

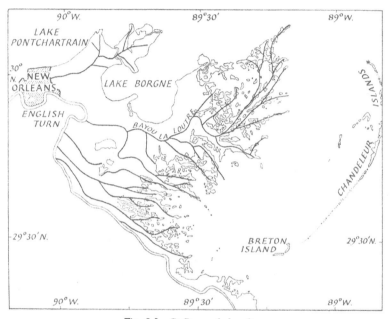

Fig. 5.3   St Bernard shoreline

final Indian occupation is the fact that natural levee crests, which stand at an elevation of about 5 m. in the vicinity of New Orleans, have been traced eastwards on the surface and in borings to a depth of nearly − 7 m. around the margins of the old delta. The gradients along St Bernard natural levee crests are five or six times steeper than those along any of today's active channels.

The general channel pattern of the St Bernard delta was more like that of the Nile than that of the bird-foot Plaquemines delta. Distributaries branched freely at places well inland. The contrast in patterns appears to be related to the fact that St Bernard channels extended gulfwards across a region not previously occupied by a

Mississippi river delta, where no large masses of backswamp clay prevented their free development.

At least half of the original area of the St Bernard delta is now submerged. Even within historic time there has been a considerable abandonment of agricultural land. Marshes have encroached across territory once used for field crops. It is distally, however, that the most radical changes have occurred.

The equivalent of the beach which is growing tangentially to the passes of the Plaquemines delta now lies about 35 km. from the irregular 'mainland' of the St Bernard area. The Chandeleur Islands, Breton Island and a line of arcuate shoals extending towards the Plaquemines delta form an offshore bar or barrier beach which is composed principally of sand and shell. Most of the shells are *Ostrea, Rangia* and other inhabitants of brackish or fresh-brackish water. Along the outer beach of the arc are important accumulations of Indian artefacts which are derived from mounds or middens located at unknown distances to the east.

The entire Chandeleur arc is moving westwards, towards the mainland, with great rapidity. A fringing 'swamp' of low mangroves along the inner side of the islands is being invaded so rapidly by the sand–shell–rubble deposit of the beach that plants in all degrees of burial occur to the west, while freshly exhumed stumps stand upright in great density over the exposed beach to the east. Blocks of fresh, unconsolidated delta clay up to 30 cm. in length appear on the outer beach during storms. A lighthouse on North Island, of the Chandeleur group, has been moved four times since A.D. 1800. Suggested rates of beach advance vary from 2 to 5 km. per century, but such values have little meaning. A single storm produces greater changes than several decades of ordinary wave erosion, and the frequency of storms is unpredictable.

Between the Chandeleur arc and the marshy mainland are sounds with depths of less than 3 m. at most places across their flat floors. Thin veneers of hard, fine sand cover the deltaic sediments beneath these bodies of water. Tidal channels, which are excavated to depths of as much as 6 m., and even more towards the islands, are generally soft-bottomed, in fine silt or clay.

The outer remnants of natural levee crests of the St Bernard mainland form 'double' islands consisting of two narrow strips of silty land which flank a channel that may contain shallow water, but which is fundamentally filled by clay, flocculants and organic debris. Some of these are several kilometres in length. Between the

old distributary channels are wide, shallow bays which are actively
encroaching upon backswamp lowlands. Water area is everywhere
gaining ascendancy over land area. Deep, meandering tidal channels
are developing rapidly across the marshes, so that their patterns are
gaining predominance over the original 'regional grain' established
by St Bernard–Mississippi distributaries.

The Terrebonne (Parish) type of shoreline (Fig. 5.4), which is
developed over a broad area to the west of the Plaquemines delta

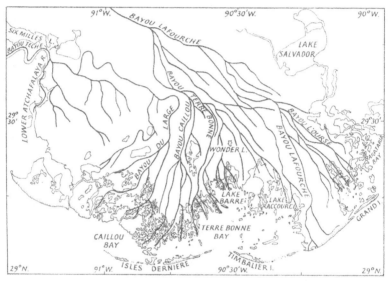

Fig. 5.4   Terrebonne shoreline

in territory generally south and south-west of New Orleans, exhibits
a stage of abandonment of alluviation that has progressed much
farther than that in the St Bernard delta. The beach equivalent to
the Chandeleur arc has reached the mainland at many places, and
in some cases has advanced inland for significant distances. The
general smoothing of coastal outlines by wave erosion is evident
everywhere. Only the inner parts of bays between the natural levees
of distributary channels remain as broad water surfaces. Most of
the old Mississippi river distributaries are represented by bands of
silt and low natural-levee ridges extending across the marshes, but
a few have maintained channels deep enough for navigation by
small fishing and shrimping boats. Bayous Lafourche, Terrebonne,
du Large and other old courses or branches of the Mississippi fan

out towards the coast in patterns that suggest a size not inferior to that of the Nile for the Terrebonne delta, and a Nile-like arrangement of channels. The Terrebonne area, however, is a complex of several old deltas, rather than a simple structure such as the St Bernard delta. Earlier Teche–Mississippi deposits towards the west were covered by later Lafourche–Mississippi deposits towards the east. To the inland are surface remnants of even older alluvial deposits.

The Lafourche, or latest of the Terrebonne deltas, reached maximum development in pre-Indian times. This does not mean great antiquity. Louisiana Indian chronology is known in great detail, but only in a relative sense. Several lines of reasoning, none resting upon very secure footing, lead to the conclusion that the oldest culture is less than 2,000 years old. Indians inhabited many parts of the Terrebonne region, but generally they lived in such places as are habitable today. No concentrations of pottery or other artefacts have been found along the beaches, which might suggest important settlements at sites now lost to the waves.

A notable feature of the Terrebonne region is the development of 'flotant' over a wide zone extending inland from the beach. This 'quaking marsh' consists of more or less firmly knit mats of living grasses and other plants anchored in a foundation of roots and debris that floats on water up to 3 m. deep. Firm silts and clays lie below the water. Early accounts tell of removing blocks of flotant and 'mining' fish from the water below. Flotant rises and falls with changes in water level. The range of lunar tides is only about 50 cm., but occasional hurricane waves raise the flotant 2 or 3 m. During some hurricanes parts of the flotant are detached and floated to new locations, where the debris deposit is locally known as 'ramassis'. Wonder Lake, a body of water over 3 km. long, was thus formed during a hurricane in 1915.

Flotant occupies the broad areas that were the original backswamp lowlands of the Terrebonne region. The inland boundary of the flotant is ordinarily a surface of 'roseau prairie', vegetation which grows on relatively firm silt or clay. The development of the flotant appears to be related to southward tilting of the Terrebonne region which has caused subsidence enough to convert former 'roseau' into places capable of growing only the semi-aquatic plants of the floating marsh. The advance of the Terrebonne beach has not been rapid enough to prevent the development of flotant.

Finally, a fourth type of shoreline extends across the western

half of Louisiana. This Cameron (Parish) type (Fig. 5.5) character-
ises land that has escaped significant Mississippi river alluviation
for possibly some twenty centuries or more. The sandy beach is
continuous and firm, with dunes up to 3 m. high along its inner
flank. While the mouths of rivers are estuarine, no bodies of water
remain to mark the bays that once existed between natural levees
of distributary streams. A beach that has moved inland at an average
rate of 200 m. per century during the last 150 years has smoothed

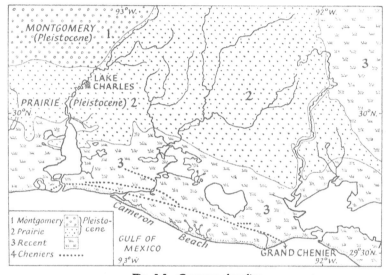

Fig. 5.5   Cameron shoreline

coastal outlines and crossed most of the delta region that once
displayed Mississippi or Red River distributary patterns. Areas
that once may have been wide flotant now lie below shallow waters
of the Gulf of Mexico. The effects of wave erosion completely
dominate the shoreline.

The records of still older shorelines are now preserved as 'cheniers'
in the Cameron region. These features are comparatively straight
beach ridges surrounded by marsh. On them grow dense stands of
oaks and other trees that emphasise their appearance as elongate
'islands' which rise above the 'sea' of the marsh, with its low grasses
and sedges.

Grand Chenier, the most recent ridge, resembles Cameron
Beach in practically all respects. It attains a maximum elevation

of about 3 m., extends longitudinally for about 100 km., and is composed principally of shells and fresh sand. Along the crest of the ridge shells are leached to some extent, and moderately corroded. Thin, dark soils have started to form, so that much of the high ground has been cleared for agriculture. At depth the shells are fresh and unleached. Many retain their original colour coatings, in spite of the fact that the distance between Grand Chenier and the existing Cameron beach is in excess of 10 km. at some places.

Older cheniers, inland from Grand Chenier, are lower, and their surficial deposits have become greatly modified by pedogenic processes. On Little Chenier and Chenier Perdu, to mention but two of the ridges, few shells remain unleached and soils are black and deep. Practically all land above high-tidal level is in agricultural use, and some lower land is protected by dikes.

Shoreline trends have not remained absolutely parallel during the period of chenier development. Older cheniers are truncated by younger, just as Grand Chenier is now terminated at either end by Cameron Beach.

Both the cheniers and Cameron Beach form obstacles which are difficult for streams to cross. The typical river from the interior follows along the inner side of each ridge for some distance before finding a transverse outlet. Marshes are ponded, so that their surfaces are wettest near the inner side of each ridge.

Most of the meandering tidal channels of the Cameron region develop as branches of the through-flowing drainage ways. Elsewhere along the Louisiana coast they lead directly to the Gulf and are prevented from entering streams from the interior by natural levees. The contrast is due partially to the effectiveness of Cameron Beach and the cheniers as barriers, so that few gulfward outlets exist, and partially to the greater age of the western marshes. The natural levees of old Mississippi and Red River distributaries have subsided below the general surface, excepting along the inner margin of the marshes, where they are perfectly distinct.

A curiosity of the Cameron region, related to the rapid inland advance of the beach, is the segmentation of several old streams by the beach. Discontinuous remnants of old channels now exist as linear lakes which are terminated at either end by beach.

The differentiation of the Cameron region into marshes separated by semi-parallel cheniers has been caused by alternations between periods marked by domination either of alluviation or wave erosion. During times when outlets of the Red and Mississippi rivers were located along the western part of the alluvial valley, or had lower courses directed westwards, a surplus of sediment collected and caused the shore to advance into the Gulf of Mexico. Diversion of the rivers to the east, caused by more or less accidental events inland, diminished the supply of available sediment, so that erosion gained dominance, and shorelines marched inland. Irregular shores undoubtedly marked periods of southward growth, while smooth shores developed with progress of inland advance. Beaches marking the farthest lines of encroachment of the Gulf across coastal Louisiana were left 'stranded' as cheniers when deltaic marshes began forming in front of them. Wave currents have great ability to form beaches and to push them inland, but are ineffective as movers of beaches in a seaward direction.

A prolonged stability of Mississippi and Red River deposition in its present position, in eastern Louisiana, could result in the advance of Cameron Beach across the entire width of the western marshes and the removal of all of the cheniers. An advance of 50 km. would bring the beach to the outcrop of the youngest Pleistocene, or Prairie, formation at practically all points. At the present rate of movement this would require about 25,000 years. If man survives that long, it is barely possible that he may desire and be able to maintain the Red and Mississippi rivers in their existing courses, but only strenuous efforts on the parts of engineers keep them there today.

Under natural conditions it is quite certain that the Mississippi would have abandoned its channel past New Orleans some years ago. The Atchafalaya river, the next natural course of the river in southern Louisiana, offers an extreme gradient advantage, a route to the Gulf about half as long as the New Orleans–Plaquemines channel. This turbulent distributary increases its discharge notably each decade, scours pools more than 30 m. deep, and now carries 25 per cent of Mississippi discharge during important floods. Its juvenile delta, which now lies inland and west of the Terrebonne coast, is growing rapidly and appears destined to become one of the world's great alluvial deposits. By the time that the Atchafalaya delta has advanced a few kilometres into the Gulf, Cameron Beach will be well on the road towards becoming a chenier.

REFERENCES

BARTON, D. C. and HICKEY, M. (1933) 'Gulf Coast geosyncline', *Bull. Amer. Assoc. of Petrol. Geol.*, XVII 1447–58 (for original naming of the geosyncline).

FISK, H. N. (1938) 'Geology of Grant and La Salle Parishes', *Louisiana Dept. Conservation, Geol. Bull.*, X (for descriptions and names of Pleistocene terraces).

—— (1939) 'Depositional terrace slopes in Louisiana', *J. Geomorph.*, II 181–200 (for terrace relationships).

—— (1940) 'Geology of Avoyelles and Rapodes Parishes', *Louisiana Dept. Conservation, Geol. Bull.*, XVIII (for naming Pleistocene formations).

—— (1944) 'Geological investigation of the alluvial valley of the Lower Mississippi River', *U.S. Army, Corps of Engrs. Mississippi River Commission (Vicksburg, Miss.)* 1 Dec.

RUSSELL, R. J. (1936) 'Physiography of Lower Mississippi River delta', *Louisianna Dept. Conservation, Geol. Bull.*, VIII 1–199 (for detail in the Plaquemines and St Bernard areas).

—— (1938) 'Quaternary surfaces in Louisiana', *Comptes Rendus du Congr. Intern. Géol.*, II 406–12 (Amsterdam).

—— (1940) 'Quaternary history of Louisiana', *Bull. Geol. Soc. Amer.*, LI 1199–1234.

—— and HOWE, H. V. (1935) 'Cheniers of southwestern Louisiana', *Geol. Rev.* XXV 449–61 (for recognition of cheniers).

6 The Erosional History of the Cliffs around Aberystwyth

ALAN WOOD

SUMMARY

The coastline from Borth to New Quay is described. Three points of general interest emerge. The coastal bevel (hog's-back cliffs, *fausses falaises*) is shown to be the remains of cliffs cut by marine erosion, and degraded by sub-aerial weathering, at several times in the past. Raised beach platforms occur at different levels, each being backed by degraded cliffs, which merge laterally with the coastal bevel. Later than these, but earlier than the spread of the latest boulder clay in this area by solifluction seawards, are steep 'fossil cliffs' cut during apparently two periods of lowered sea-level. Where the beach platform is cut in solid rocks its outer edge is coincident with the course of one or other of these fossil cliffs.

INTRODUCTION

THE coast around Aberystwyth is particularly suitable for morphological analysis, for rocks of almost identical lithological type occur for 30 miles from Borth to beyond New Quay. These rocks, shown in Plate 5, are a typical greywacke suite, consisting of a rapid alternation of fine-grained badly sorted sandstone, or siltstone (greywacke) beds with darker, rudely cleaved mudstones. They belong to a local lithological facies of Upper Llandoverian age, the Aberystwyth Grits. These rocks, resistant and uniform in character, have preserved the impress of early erosional episodes, which in areas of softer rocks would have long since disappeared. Beyond this, lithology has had little influence on the cliff form, the resistance to weathering is roughly uniform and any difference of cliff slope or height is due either to geological structure or to a variation in the erosional history of the cliff face.

The coastline (Fig. 6.1) is gently curved, and is evolving under generally uniform conditions. The direction of greatest fetch is southwestward, and the greatest and most effective storms are from the south-west. Consequently there is a northerly drift of beach material, which has pressed rivers such as the Dyfi and the Ystwyth to the northern side of their estuaries.

Structurally, the beds have been folded into a series of periclinal

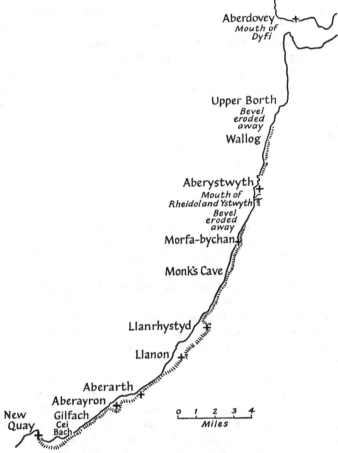

Fig. 6.1 *The coastline from Borth to New Quay, showing embayments in the coastal bevel*

folds, *en échelon* in depth as well as laterally, and the folds tend to run along the gentle curve of the coast. Joints in the hard greywacke beds appear to have considerable effect on the form of the cliffs. When the dip is inland the cliffs tend to slope seawards at a uniform angle, determined by the seaward dip of the joints, while when the beds are horizontal the cliffs may be almost vertical. Seaward-sloping beds display a landslip topography, and the cliff face slopes at much the same angle as the dip of the beds. Clearly features due in this way to the attitude of the beds, and others due to localised weaknesses such as faults, thrusts or shatter belts, can be easily picked out, and thus structural influences can be recognised, given their due importance and thereafter ignored. The erosional history is then the only unknown factor in the shaping of the cliffs.

The heights of platforms mentioned in this paper have been determined by aneroid in the case of some of the higher ones, and by direct triangulation in the case of platforms visible on the faces of cliffs. Levels obtained in this way were referred to a local high-water mark and this in turn to O.D. On some of the wilder beaches the levels may be inaccurate by reason of miscorrelation of high-water marks. It is believed that the errors are not significant in view of the variation in level observed in recent beach platforms.

PREVIOUS WORK IN THE AREA

While mapping the area between Cardigan and Aberystwyth, Ramsay (1846) was struck by the general seaward inclination of the land surface above the line of cliffs, stating that this great plain was "bounded by distant inland hills, easily comparable to many an existing line of lofty coast'. Keeping (1881) extended Ramsay's observations, recognising the existence of two denudation surfaces in Cardiganshire, a concept elaborated by Jones (1911), who named them the High Plateau and the Coastal Plateau. The latter forms the Cliff-top Surface of the present paper, and is the surface which kindled Ramsay's interest.

The earliest account of the cliffs themselves was by Mellarde Reade (1897). He gave an account of the glacial deposits as seen in cliff sections, and first observed the 'pre-glacial' cliff to the south of Aberystwyth, showing how the boulder clay dipped off the nearly vertical rock face. Challinor, in a series of papers (1931, 1948, 1949), has described the coastline of this district, laying special stress on the development of a 'coastal slope' above the line of cliffs. This he considered to be a normal development due to accentuation of sub-aerial weathering above the rapidly eroded, nearly vertical cliff of modern marine abrasion. The pre-glacial nature of the coastal slope, south of Llanrhystyd and south of Aberarth, was recognised by Challinor. The well-known book by Steers (1946) gives an interesting compilation of facts and discussion of the cliffs. Cotton (1951) reinterpreted one of Challinor's photographs of the coastline near Wallog, claiming that the coastal slope was an ancient cliff, of inter-glacial age, graded by sub-aerial agencies. This view had been independently arrived at by the present author (1952).

GENERALITIES

From the foregoing account, it will be seen that the problem which has aroused most discussion in the past is the origin of the coastal bevel (= coastal slope). A puzzling feature, not previously adequately discussed, is the varying lower limit of this bevel. In the present district it may extend to sea-level, as on the slopes of Pen Dinas, or be restricted to the tops of the cliffs, as between Aberystwyth and Clarach. Furthermore, it may sometimes be separated from the sea by a considerable stretch of boulder clay, as between Aberarth and Aberayron. The bevel is not restricted to this stretch of coast; it is found in North Wales on the resistant rocks of the Lower Cambrian of Hell's Mouth Bay, and in South Wales, on the Pre-Cambrian cliffs near Littlehaven, and at many other places around the Welsh coastline. In Devon and Cornwall the bevel forms the hog's-back cliffs of Arber (1911), and Guilcher (1954) in a treatise on coastal geomorphology states that this feature, called by him *fausse falaise*, is the normal profile in cliffs cut in hard Palaeozoic rocks everywhere on the Atlantic coasts. As to its origin, Challinor's views have already been given. Arber (1911) considered the hog's-back cliffs to be 'the product of aerial, and not marine, denudation', largely because similar slopes could be observed far inland. Wilson (1952) thought that the bevel was largely, if not entirely, a fault-line feature, while Cotton (1951) considered the coastal bevel in the present area to be a degraded sea cliff. On the other hand Guilcher (1954) claims that it is formed by sub-aerial agencies alone, being 'un versant à modèle uniquement continental'.

The second great problem in the area, not discussed since 1897 when Mellarde Reade first noticed it, is the age and origin of the pre-boulder-clay cliff, and its relationship to other surfaces.

Finally, the splendid cliffs, often descending sheer several hundred feet into the sea, gave a vivid impression of the power of marine erosion. As O. T. Jones (1924) has shown, the traditional British view that unconformities are former plains of marine denudation probably owes its existence to Rámsay's early impressions in this area of lofty cliffs and wide beach platforms. How far these fresh cliffs actually owe their appearance to post-glacial marine erosion is a point that has never been discussed, and one on which the present study sheds some light.

OBSERVATIONS ON THE COAST

(a) *The coastline between Morfa-bychan and Mynachdy'r-graig, four miles south of Aberystwyth*

This area, part of which is shown in Fig. 6.2, gives an exceptionally clear picture of the relationships of the cliff forms. The prominent seaward-sloping cliff visible to the south of Aberystwyth is determined by a seaward dip of the beds; beyond it a raised beach platform is visible in the distance, first appearing 800 yds NNE. by N. of the farm Morfa-bychan. The landward edge of the platform is determined by the junction of its gently sloping surface with the steeper slope of the coastal bevel. Solifluction-distributed boulder clay (Head) everywhere covers the surface, and determines the elegant concave slope running up into the bevel behind. The actual junction of platform and bevel is not seen in section in the cliffs, partly owing to slipping and partly because of the possible presence of a remnant of a higher platform. However, the first exposure of the solid rock surface, in a vertical cliff, is not far from this junction. Here the solid rock is covered by 17 ft of Head, and the rock surface stands at 42 ft O.D. The nip cut by present-day wave attack at the south end of the cliff, by the old lime kiln, is 8 ft above O.D., while at the north end it is only 2·5 ft above this level. A wide beach platform is present. When the rocks are traced southwards they are found to vanish behind a mantle of boulder clay; simultaneously the beach platform narrows and vanishes. Boulder-clay cliffs extend southwards for 700 yds, the occasional presence of large angular blocks of the greywacke series at the cliff foot, or a trace of solid rock high up the face, showing that the boulder clay is a relatively thin veneer. This is quite clearly proved south-west of Morfa-bychan, where a solid rock cliff, nearly 100 ft high, suddenly emerges from behind the boulder-clay cliff (Plate 5B). The slope of the just exhumed rock face is 80°, that of the recent wave-cut cliff in the same beds 83°–90°. The boulder clay in contact with this 'fossil cliff' dips steeply off it, as though the material had slipped over the cliff edge and accumulated as a talus heap at its foot (Plate 5A). At the other end of the small embayment the fossil cliff is again buried by boulder clay.

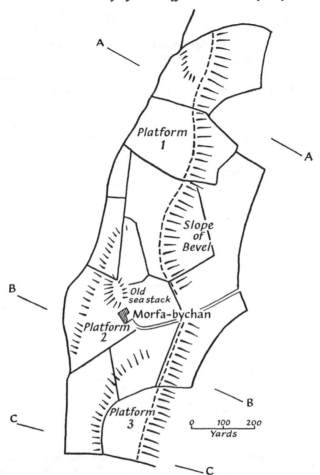

Fig. 6.2 Detail of platforms at Morfa-bychan. The lines of the cross-sections shown in Fig. 6.3 are indicated.

A striking feature here is that a beach platform cut in solid rock is only developed where the cliffs are composed of rock, and that its outer margin is a prolongation of the line connecting the junction of the fossil cliff and boulder clay to north and south. This is good evidence that the buried cliff was cut by a sea whose level was below that of the present one. The outer edge of the beach platform is determined by this cliffline.

There is an unmistakable difference in height between the rock platform at this locality and that first described, which is reflected in the form of the platform surface above. As shown in Fig. 6.3 there is a rise in the boulder-clay-mantled surface as this higher platform is approached. In this one locality, therefore, we have sure evidence of the existence of two platforms, their actual rock surface being exposed in the cliffs, and we can observe how the boulder-clay spread

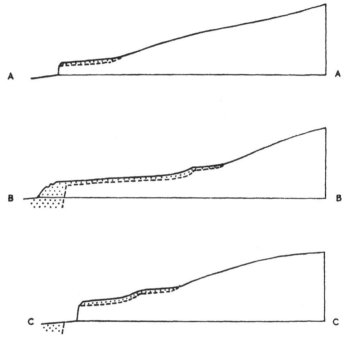

Fig. 6.3   Cross-sections of platforms at Morfa-bychan, showing their relationships with the boulder-clay veneer. Vertical and horizontal scales approximately equal. Boulder clay with dots, solid rock without ornament.

is mantled over and moulded upon their surfaces. Furthermore, there appears to be a third, still higher, platform to the south of Morfa-bychan, whose outer edge slopes steeply to the flatter ground immediately above the high fossil cliff. The solid rock surface of this platform is nowhere exposed, but probably lies at about 180 ft O.D. The coastal bevel is rather dissected here, perhaps owing to the seaward dip, but it runs down below the 100-ft contour above the lowest platform, while its lower edge is just on the 200-ft contour above the highest platform south of Morfa-bychan.

From this spot southwards the cliffs are composed of boulder clay for nearly a mile, with the fossil cliff everywhere just behind. The surface of the narrow platform above is boulder-clay mantled, cultivated and backed by the coastal bevel. South of the farm Ffos-las the surface of the boulder-clay platform declines, presumably passing over the junction between a higher and lower rock platform. A tiny outlier of the higher platform (an old sea stack) is exposed at the edge of the cliff between Ffos-las and Mynachdy'r-graig, and another, larger one with a good reversed slope landwards lies just seaward of the latter farm. The solid rock surface of this stack is at 132 ft O.D., the rock surfaces on either side are below 100 ft. At a point 250 yds south of Monk's Cave the solid rock surface visible in the cliff is at 84 ft O.D., and at the south end of the section the height of the junction of solid rock platform and coastal bevel is 96 ft O.D. Here the coastal bevel continues southwards, capping freshly cut cliffs, and it is clear that marine erosion in recent times has removed the traces of all these former raised beach platforms.

Though the lower platforms have been thus destroyed, traces of still higher platforms, at about 315 ft O.D., occur half a mile south of Monk's Cave, perched high above the bevel, which here slopes steeply seawards. Behind the two small, nearly flat remnants the coastal bevel is slightly dissected; nevertheless its seaward slope is in places as high as 30°. The outer edge of the platforms is abruptly cut off by the lower section of the coastal bevel, which here slopes at 32° and steepens still further seawards, till it is sloping at over 40° seawards at its junction with the Recent cliffs.

A particularly significant section occurs above an unnamed headland 600 yds south-west of this point, where the coastal bevel runs down to 25 ft O.D. before being cut off by Recent erosion. This must mean that at some period in the past any traces of the three platforms recognised to the north were eroded away, prior to the formation of the lower part of the coastal bevel.

(b) *Section north of Aberarth*

The steep cliffs of this section appear to have a simple history when viewed from afar (Plate 5F). A coastal bevel, sloping at 32°, is transected by high, nearly vertical sea cliffs. This simplicity is, however, completely illusory. Going northwards along the coast from Aberarth, steep boulder-clay cliffs occur, and this material is found to rest on, and be banked up against, a rock surface sloping at 45° seawards. Fine earth pillars have been formed by water

ACG D 2

running down the solid rock slope, and dissecting the boulder clay lying on it. The rock surface emerges from below boulder clay obliquely to the present coastline, so that further to the north Recent erosion has cut into it. Here a narrow beach platform, with its outer edge coincident with the prolongation of the sloping rock surface, is seen. The strike of the beds is nearly parallel to the coast and they dip at 27° inland. The Recent cliff has a seaward slope of 77°–88°, in strong contrast with the pre-boulder-clay surface cut in the same beds. Above there is a trace of a platform at about 95 ft O.D., in the shape of a concavity at the base of the coastal bevel; towards Aberarth this has been modified by fluviatile erosion at various dates. The fossil cliff slope between the platform above and the modern cliff below extends northwards for 200 yds, sloping quite uniformly seawards, and the rock platform at beach level widens as the Recent cliffs become higher. The cliffs then turn slightly eastwards and the modern cliff becomes continually lower, and the beach platform narrower, till a boulder-clay-mantled fossil cliff descends nearly to sea-level. Here the fossil surface dips at 55° seawards. Recent cliffs in the same beds are sloping at 75°. The contrast in slope between this buried cliff and the first is striking.

At the north end of the Aberarth section a relatively low fossil cliff can be seen running behind boulder clay, below the bevel which turns inland at this point. The angle of slope of the fossil cliff is 65°–68°, that of the modern cliff 70°, steepening to 80° in a wave-cut notch at the base of the cliff where thicker beds of mudstone occur. The beach platform widens southwards, as Recent erosion has been more effective, and the cliffs become higher. The remains of two raised beach platforms can be recognised on the top of the cliff, by the fact that the lower edge of the coastal bevel becomes concave upwards. The lower one is at 88 ft O.D.; above it the ground rises rapidly to the higher platform, which runs along the cliff top southward at 185 ft O.D. The trace of this platform is removed by Recent erosion before the headland is reached, and at the headland itself the bevel runs down to a concavity at 103 ft O.D. This appears to indicate that the platform at 185 ft O.D. had been removed by erosion at this lower level.

(c) The Boulder-Clay platform between Llanrhystyd and Llanon

The coastal bevel curves inland five times along this coast, and comes to lie behind a platform of boulder clay gently sloping seaward (Plate 5D). The northern part of the present embayment

shows features which are, *mutatis mutandis,* common to them all, and
have importance as far as the later history of the coastline is concerned. North of Llanrhystyd the bevel swings inland on the north
side of the valley of the Wyre, and merges with slopes of sub-aerial
denudation. To the south-west of Llanrhystyd, and as far as Llancn,
the coastal bevel is prominent in the landscape. Its lower margin
lies first at 50 ft O.D.; below it the surface of the boulder-clay
platform declines seawards to below high-water mark. A storm
beach separates the cultivated fields from the sea. The base of the
bevel rises to above 100 ft as one approaches Llanon, as does the
surface of the boulder-clay platform. The rise in the latter can be
appreciated as one travels along the main road, and it appears to
take place in several steps. This is doubtless due to one or more
buried rock platforms, across which the boulder clay has been
plastered. The seaward edge of a buried platform can be observed
from the seashore as an almost cliff-like descent which runs from
behind the old lime kilns towards Llanon and then curves round
into the valley of the Peris. Solid rock is nowhere seen in the foreshore or cliffs, but the cliffs are highest (42 ft O.D.) where the buried
platform is closest to the shore. The gentle uniform slope of the
surface of the whole boulder-clay platform is particularly striking.
Exposures in the cliffs north of Llanon show clearly that the smooth
surface, here at least, is due to infilling of hollows in boulder clay
by water-washed, probably solifluction-spread, debris.

(d) The remainder of the coastline

The Coastal Plateau is particularly clearly developed above Aberdovey. Seen from Ynyslas the cliff-line at its upper limit is striking,
especially in winter when picked out by snow. The slope of the
Plateau below, cut in solid rock, is $4\frac{1}{2}°$ seawards. It terminates in
the fine grass-covered and gently ravined old cliff-line to the north
of Aberdovey, separated from the sea by sandhills.

To the south, from Ynyslas to Upper Borth, a shingle spit,
testifying to the reality of a northward drift of beach material, has
pushed the river Dyfi northwards. Behind it lie the deposits of Borth
Bog. O. T. Jones (1916) has discussed the history of this stretch of
coast, of which a good account is also given by Steers (1946).

From Upper Borth to Wallog cliffs occur, rising to nearly 400 ft
south of Craig y Delyn. At first the cliffs are low, and nearly vertical,
the picture being blurred by traces of glacial and fluviatile erosion
as far as Borth Monument. Below the latter a cliff 200 ft high slopes

seawards at a uniform angle of 48°. The strike of the rocks is slightly oblique to the cliff face, and the overturned beds dip at 65°–70° landwards. A broad beach platform exists, with remnants of stacks standing some 8–10 ft above its surface, and one large one rising to 18 ft. Between this large stack and the foot of the cliff a zone of the beach platform nearly 30 ft wide is nearly 1 ft higher than the platform to either side. Beyond the Monument a remnant of a valley, sloping seawards and floored with a veneer of Head, occurs, before the narrow re-excavated valley of the Wennal is reached. The cliff tops here and as far as Craig y Delyn are clearly the product of sub-aerial erosion, and the cliffs likewise have been shaped by Recent erosion.

Beyond Craig y Delyn the higher cliff tops are spurs of the Coastal Plateau. The coastal bevel is probably represented by a tiny seaward convexity above Craig y Delyn itself, and on the top of the next headland but one to the south, but the best development is above the cliffs to the west of Moel Cerni. Here the bevel is entirely grass-covered and slopes at 33° seawards, while the rock cliff below has a slope of 63° in the same direction, the rocks themselves dipping landwards. Above an unnamed headland 570 yds WNW. of Moel Cerni the coastal bevel shows a concave slope at its base. This tiny surface has a rock level of 180 ft O.D., and its junction with the continuous bevel to the north is obscured by a shallow valley.

On approaching Wallog the bevel swings inland above an old platform cut in solid rock and mantled with solifluction head. From the beach this platform can be seen to have had a complex history. The junction of solid rock with the slope of the bevel is 70 ft O.D., but south of this the rock surface rapidly descends to 30 ft O.D. and two boulder-clay-filled gashes descend to high-water mark (say 12 ft O.D.). Nearer Wallog itself the rock surface maintains a fairly constant level of 44 ft. The surface of the fields above slopes gently and uniformly seawards above all these irregularities. Wallog stream falls to the sea over a rock barrier originally about 25 ft O.D. Beyond the stream, below Wallog House, is a cliff in boulder clay, concealing an old cliff in solid rock. At the contact of boulder clay and old rock cliff angular blocks are common, embedded in boulder clay. The line of the old cliff is oblique to the present coast, as is proved by the widening of the rocky beach platform in a southward direction as far as Clarach. At one headland, south of Wallog, the grass- and boulder-clay-covered slopes descend seawards on a rock surface sloping at 43°, which cuts across the rock structure.

Above the present cliff top is the bevel, broken down into a number of old platforms, well shown in Plate 17 of Steers (1946).

Between Clarach and Aberystwyth is the type of area of the coastal slope of Challinor. The bevel, sloping at angles up to 34°, occurs high up above the precipitous modern cliffs (Plate 5E). No traces of platforms are seen, except for a concave slope to the base of the bevel, visible from the north end of the Promenade at Aberystwyth.

The erosional history at Aberystwyth itself is extremely complicated. The bevel swings behind North Road, and exists below the National Library, which is itself built in a rock platform at 185–90 ft O.D. The grassy slope descending south-west of the National Library may represent a surface cut in a period of falling sea-level with a small sea stack, partially quarried away, at its outer margin, and the Castle appears to be on another platform at 58 ft O.D. These and many other features must have been affected by glacial erosion and by fluviatile erosion at several levels. The same is true of the apparent platforms on the north-west slopes of Pen Dinas, and the possible sea stack near the mouth of the Ystwyth. The seaward slopes of Pen Dinas probably represent the coastal bevel, here modified by fluviatile erosion, and descending below sea-level.

The erosional history of the next section of the coast has been described in section (*a*). Southwards of the point where the bevel descends to 25 ft O.D. recently cut cliffs, capped by a prominent steeply sloping coastal bevel, continue towards Llanrhystyd. On the headlands the bevel extends seawards, often running down to the 100-ft contour, and once extending below it. To the north-west of Llandrhystyd the bevel turns inland and its relationship with the boulder clay lying seaward has been described in section (*c*).

Immediately south of Llanon the lower edge of the coastal bevel is at 50 ft O.D., and the boulder clay slopes gently seawards, ending in low cliffs, but within a quarter of a mile the base of the bevel has risen 50 ft, presumably over a remnant of buried rock platform, and the coastal boulder-clay cliffs are higher in consequence. Beyond Morfa Mawr the cliffs have been described in section (*b*).

At Aberath itself a platform of rock is visible by the bridge, in the stream bed of the Arth, at a level of 24 ft O.D. The platform slopes seawards, its outer edge being at about 19 ft O.D. It is clearly of pre-boulder-clay age. South of this point solid rock is not seen again in the cliffs till beyond Aberayron. The coastal bevel lies behind a boulder-clay spread, again with a probable buried platform as Aberayron is approached. Near Tre-newydd-fach the boulder-clay surface falls below high-water mark, and at this point, as near Llanrhystyd, no coastal erosion has occurred since the sea attained its present level.

The coastal bevel swings seawards at the headland, Pen y Gloyn, south of Aberayron, the uniform height of the rock cliff here being due to the parallelism of bevel and coast, not to the presence of a raised beach. Once the headland is passed the bevel must have once swung round through a right-angle, to run nearly parallel to the coast, for the beach platform narrows south-westwards. Evidence of pre-boulder-clay erosion at 12–14 ft O.D. is found 400 yds to the south-west of the headland, just before the sold rock goes behind the boulder-clay cliff. For nearly a quarter of a mile the sea beats against boulder clay, which slips down from the steep (32°) coastal bevel behind. A break of slope at the base of the bevel is probably the summit of the fossil cliff, against which the boulder clay is plastered. This break of slope merges south-westwards into a boulder-clay flat above a solid rock platform 48 ft above O.D. The seaward edge of the solid rock platform slopes steeply seawards, at 45°, this grass-covered slope being the trace of a degraded fossil cliff, cut in solid rock and with a thin veneer of boulder clay. Lower cliffs, nearly vertical, the product of Recent marine erosion, cut into this slope on the seaward side. Both north and south of this main platform there is a suggestion of a platform at a lower level. Beyond Gilfach the coastal bevel is grandly seen, sloping seawards above the high castellated cliffs. The beach platform is narrow, its outer edge marking the line of the fossil cliff. This comes to coincide with the modern cliff-line at the north-west end of Little Quay Bay, and the magnificent cliffs here are almost entirely the product of interglacial erosion and have been recently exhumed and slightly freshened by Recent marine action. These fossil cliffs, capped by the bevel, run behind boulder clay in Little Quay Bay, in a completely convincing exposure, and it can be seen that the slope of the fossil cliffs (68°) is echoed again and again by cliffs in the vicinity. The wave-cut base of the modern cliffs, however, is steeper, sloping at 75°–80°. Behind the boulder clay the coastal bevel occurs, presumably often sharpened at its base by the fossil cliff. It curves round behind the boulder clay of Little Quay Bay and New Quay Bay to form the slope behind New Quay itself. The first exposure of solid rock appearing from below the boulder clay is east-south-east of New Quay. Here the rock surface emerges obliquely to the shore-line, sloping at 45°, and above it pebbles in the boulder clay are aligned parallel to each other, and dip at 32° in the same direction as the surface below. It is noticeable that the rocky beach platform widens as the rock cliff face becomes higher, and in several places the slope, if prolonged, would meet the outer edge of the platform.

DISCUSSION OF RESULTS

(a) Raised Beaches

The rock benches described on p. 103 slope gently seawards, have relatively plane surfaces, and are not in regions where they could have been cut by fluviatile erosion. There can be no doubt that these platforms are raised beaches, despite the lack of beach deposits on them, and the coastal bevel behind is clearly a degraded cliff. It seems likely that wherever the coastal bevel ends seawards in a concave slope, as seen at many places along the coast, a raised beach was formerly present. Former beach platforms have been recognised at 315, 185, 88–95, 70, 45–8 and perhaps 20 ft above sea-level, and it is clear that each has been attacked and in large part removed by erosion at a series of successively lower sea-levels. Their preservation, even as remnants, can be attributed to slight differences in the angle of wave attack and in the submarine accumulation of debris, guided in part by the distribution of more massive beds, and especially by the course of oblique folds and faults. As sea-level fell, these oblique structures caused headlands and bays to migrate laterally, varying the locus of greatest erosion, which itself set up a further train of connected events.

When no raised beaches occur, and the coastal bevel is prolonged nearly to sea-level, or below sea-level, as on the seaward slope of Pen Dinas or along the courses of the three southernmost embayments, erosion at a lower level has clearly removed them. The traces of this widespread erosion are discussed in the next section.

Attempts by geophysical means to ascertain the depth of the submerged beaches associated with these cliffs have failed.

(b) The fossil cliffs and the submerged beaches

The fossil cliff which runs behind boulder clay near Morfa-bychan (p. 102) is to the eye as fresh as the Recent wave-cut cliffs. Actual measurement shows that it is sloping slightly less steeply. The same is true of cliffs at the northern end of the Aberarth section, and those at Cei Bach. Only a short period of time had elapsed between the withdrawal of the sea and the smothering of these surfaces by boulder clay. On the other hand cliffs below the boulder clay at the southern end of the Aberarth section, and those near New Quay, are much less steep than the Recent cliffs in the same rock, as are the remnants of cliffs south of Wallog and between Aberayron and Gilfach. In all these cases, however, the fossil cliffs, prolonged, meet the edge of the present-day rocky beach platform. Since the second set of cliffs suggests that a considerable period of sub-aerial erosion had occurred before final burial by boulder clay, while the other cliffs are nearly fresh, it would seem that low-level erosion occurred at two considerably separated periods. There is some evidence at the south end of the Aberarth section that one set of cliffs intersects the other, as there is between Aberayron and Gilfach. It may tentatively be assumed, therefore, that two separate periods of low-level marine erosion occurred, both ante-dating the local drift.

(c) *The coastal bevel*

Previously it has been shown that the coastal bevel behind the raised beaches must be a degraded cliff. The slopes behind raised beaches of different heights run together laterally and appear as a slope of uniform inclination above the lowest beach platform. This slope, the coastal bevel, must therefore have been cut at different times. Furthermore, where a platform notches the bevel, as south of Monk's Cave, the part behind and that in front of the platform must be of different ages. Finally, where platforms are absent and the coastal bevel is prolonged nearly to sea-level, it must be of even younger age. It seems that the coastal bevel is a polygenetic thing, a constant slope to which cliffs become degraded. Though it was often cut into, and rejuvenated, by marine erosion, the period of time between each attack was so long that the cliff was repeatedly reduced to a uniform slope by sub-aerial erosion, possibly under periglacial conditions. Presumably the development of a grass cover when the rocks weather back to the angle of repose retards erosion so much that the remains of cliffs of different ages can run together. The bevel generally has a seaward slope of 32°–33°, the steepening of its base south of Monk's Cave being probably due to the presence of a remnant of the earlier of the two fossil cliffs.

(d) *Boulder-clay platforms*

At first sight the boulder-clay-mantled platforms in the three embayments, sloping gently seawards, might be interpreted as raised beaches of Post-glacial age. However, as has been shown, they split up into a number of terraces, and it would be necessary to invoke a number of periods of Post-glacial erosion at different levels. The evidence given on pp. 106 et seq. shows that these terraces are moulded on solid rock platforms below, and it must be concluded that the boulder clay was spread seawards, below the bevel, by solifluction at the end of the last glacial period.

(e) The extent of Post-glacial marine erosion

At the end of the ice age broad stretches of boulder clay, whose surfaces were made even by solifluction, sloped gently to the sea, beyond the bevel and the fossil cliffs. It has been shown (pp. 103–4) that the boulder-clay surface was higher where it rested on solid rock platforms, or spread from bold cliffs, so that the coastline had a strongly digitate margin. The Post-glacial rise in sea-level continually renewed the attack of the sea on these boulder-clay spreads, and by a combination of erosion and drowning of the margins the sea advanced inland probably several miles. However, if the present slopes of the boulder clay are projected seawards, the amount of erosion in stiff boulder clay since the sea attained its present level (say 5,000–6,000 years ago) is only a few hundred yards. In places no erosion has occurred at all (pp. 109–10.). As solid rock was revealed, the waves attacked it, and the amount of erosion may be measured by the width of the rocky beach platform; this may even be absent where the fossil and the recent cliff coincide, but an average width is 250 ft. The rate of erosion of solid rock is therefore small, perhaps half an inch a year. The volume of material removed by erosion is likewise smaller than might be thought; at many places the bevel, where prolonged, meets the outer edge of the wave-cut platform, so that only a wedge of rock has been removed; at other places low-level platforms existed seaward of the present cliff (as perhaps at Borth) and the task of the sea was greatly reduced.

(f) Comparison with other areas

Considerable caution is necessary in attempting to correlate beaches laterally from this rather isolated area. It can certainly be said that the 'Post-older-drift Platform' of Gower is probably similar to the boulder-clay platforms discussed on pp. 103–4, and thus is not a beach at all. George (1932) had recognised this possibility; later writers were not so cautious. The Coastal Plateau may well correlate with the 430-ft Platform of Devon and Cornwall, though the widespread 200-ft platform of that area is, surprisingly, absent unless it is represented by the 185-ft platform of which only a few traces are seen. The lower platforms are difficult to correlate; they are certainly considerably earlier than the latest glaciation of this region, as is shown by the degradation of their cliff-line as compared with the freshness of the fossil cliffs. Though detailed correlations are not made, this might be possible over surprisingly great distances. This is a region of contraposed shorelines, and the pictures accompanying Clapp's paper (1913) first defining contraposed shorelines suggest that a fossil cliff and raised pre-boulder-clay beaches might be recognised in British Columbia also. Furthermore, the general sequence of former sea-levels here recorded corresponds quite closely with the Mediterranean levels cited by Zeuner (1948).

Finally, it should perhaps be said that although the simplest explanation of the sequence of events has been adopted here, a number of pieces of evidence suggest that the actual history may have been more complicated.

REFERENCES

ARBER, E. A. N. (1911) *The Coast Scenery of North Devon.*

BROWN, E. H. (1952) 'The River Ystwyth, Cardiganshire: a geomorphological analysis', *Proc. Geol. Assoc.*, LXIII 244.

CHALLINOR, J. (1931) 'Some coastal features of North Cardiganshire', *Geol. Mag.*, LXVIII 111.

—— (1948) 'A note on convex erosion-slopes, with special reference to North Cardiganshire', *Geography*, XXXIII 27.

—— (1949) 'A principle in coastal geomorphology', *Geography*, XXXIV 212.

CLAPP, C. H. (1913) 'Contraposed shorelines', *J. Geol.*, XXI 537.

COTTON, C. A. (1951) 'Atlantic gulfs, estuaries, and cliffs', *Geol. Mag.*, LXXXVIII 113.

GEORGE, T. N. (1932) 'The Quaternary beaches of Gower', *Proc. Geol. Assoc.*, XLIII 291.

GODWIN, H., SUGGATE, R. P., and WILLIS, E. H. (1958) 'Radiocarbon dating of the eustatic rise in ocean-level', *Nature*, CLXXXI 1518.

GUILCHER, A. (1954) 'Morphologie littorale et sous-marine' (Paris: Presses Universitaires de France).

JONES, O. T. (1911) 'The physical features and geology of Central Wales', *Aberystwyth and District* (National Union of Teachers souvenir), 25.

—— (1924) 'The Upper Towy drainage system', *Quart. J. Geol. Soc.*, LXXX 56.

——, in JOHNS, D., JONES, O. T., and YAPP, R. H. (1916) 'The salt marshes of the Dovey estuary', *J. Ecology*, IV 27.

KEEPING, W. (1881) 'The geology of Central Wales, with an appendix on some new species of Cladophora by Charles Lapworth', *Quart. J. Geol. Soc.*, XXXVII 141.

RAMSAY, A. C. (1846) 'On the denudation of South Wales and the adjacent counties of England', *Mem. Geol. Surv.*, I 297.

READE, T. M. (1897) 'Notes on the drift of the Mid-Wales coast', *Proc. Liverpool Geol. Soc.*, VII 410.

STEERS, J. A. (1946) *The Coastline of England and Wales* (Cambridge University Press); 2nd ed., 1964.

WILSON, G. (1952) 'The influence of rock structures on coastline and cliff development around Tintagel, North Cornwall', *Proc. Geol. Assoc.*, LXIII 20.

WOOD, A. (1952) 'Cliffs in Cardigan Bay and the extent of Post-glacial marine erosion', *Geol. Assoc. Circular*, 542.

ZEUNER, F. E. (1948) 'The lower boundary of the Pleistocene', *Rept of Proc. 18th Int. Geol. Congr.* (London), Pt 9, 126.

7 Land Loss at Holderness[1]

HARTMUT VALENTIN

ALONG the coast of the North Sea, between Flamborough Head and Kilnsea, the low hills of Holderness reach the coast and form cliffs. These hills are composed of ground moraine, terminal moraine and fluvio-glacial deposits; they make a 61·5-km. stretch of cliffs which rises to a maximum height of 35 m. From the crest of these cliffs it is immediately apparent that the land is retreating under the attack of the sea. At one point bunkers built on the top of the cliff in 1940–1 have tumbled to the beach and are now piles of rubble; in the same place blocks of concrete that had been placed at the foot of the cliff are now seen many metres to seaward. At another place a coast road that is shown on the latest map as passable is now interrupted. Everywhere the bare cliffs, devoid of vegetation, bear witness to the continuing destruction by the sea; indeed, this is a textbook example (Valentin, 1952, p. 57).

Three problems may be noted: first, how quickly have the cliffs retreated in the recent past; second, what causes this loss of land; third, can the cliff recession be stopped? The present study is divided into three parts: (1) measurement of the rate of recession; (2) explanation of the recession; (3) the fight against the loss of land.

I. MEASUREMENT OF THE RATE OF LOSS OF LAND

Several attempts have been made to assess quantitatively the loss of land in Holderness. A good review of the early work is provided by Reid (1885, p. 94) and the results of later studies were collected by Sheppard (1906). The evidence that was available at the turn of the century was evaluated in theleng thy reports of the Commission on Coast Erosion (1907–1911). Hence it is unnecessary to examine this early material, and we may turn to the notable contribution of Thompson (1923), who, in common with other workers, concentrated on limited sections of coast and was concerned with the loss of land over relatively short periods of time. We lack evidence for a large number of points on a long stretch of coast and for a long

[1] This paper was first published in a *Festschrift* for Otto Quelle.

period of time. The present paper attempts to fill a little of this gap.

During field studies of the glacial morphology of Holderness in 1952 (Valentin, 1954c), the author used a 100-ft tape to measure the distance of the cliff top from near-by topographical features, such as old houses, footpath junctions, hedges and ditches. Most of these reference points were later identified on the earliest 6-in. maps of the Ordnance Survey (for the year 1852) housed in the library of the University of Cambridge. On account of the large scale of these maps (1:10,560) it was possible to measure the distance from the reference points to the top of the cliff to within 10 ft (equivalent to 3 m.). The difference between the 1852 and 1952 measurements gives the loss of land over the period of one hundred years, within an error of ± 3 m.

Since some of the military installations were still in use, it was not always possible to gain acess to the coast, and in these instances erosion since 1852 was obtained by comparing the 1852 map with the most recent O.S. 1:10,560 map, which was correct for 1951. Considering that measurement on the map is accurate only to 3 m., the margin of error when map evidence is used for both years is ±6 m., and hence these data cannot be considered to be very reliable.

Altogether 307 observations were obtained at intervals of approximately 200 m. along the 61·5-km. stretch of coast. It was impossible to space the observations at precisely regular intervals of 200 m. owing to the difficulty in finding suitable topographic reference points. Nevertheless, the data meet the requirements of close spacing of records along the coast and are for a long period of time – one hundred years.

The results obtained from these measurements are shown in Table 7.1 (arranged from north to south). Lack of space precludes a detailed description of the locations of the points at which the measurements were made, but the National Grid provides a sufficiently accurate means of identification, and six-figure grid references are included in the table. The observations are grouped by parishes, with the mean value for each parish shown at the foot of the relevant column of figures. For international use, it is convenient to express the rate of land loss in metres per annum (Valentin, 1952, 1953a, 1954a, b).

The parish mean values are shown in Table 7.2. By multiplying the annual recession by the length of coast in each parish, the annual loss in square metres can be estimated. Furthermore, one can calculate the volume of material lost in each parish if the area of land eroded away is multiplied by the average height of the cliffs. However, it is difficult to estimate the height of the cliffs owing to

Hartmut Valentin

TABLE 7.1

Land Loss (–) and Gain (+) from 1852 to 1952

Grid reference	metres p.a.	Grid reference	metres p.a.	Grid reference	metres p.a.
198 684	−0·18	170 629	+0·08	178 563	−1·56
196 682	−0·18	170 627	+0·08	179 562	−1·65
195 680	−0·24	170 625	+0·31	179 560	−1·53
194 679	−0·18	170 623	+0·23	180 557	−1·65
192 676	−0·09	170 620	0·00	181 556	−1·62
191 674	0·00	170 618	0·00	181 555	−1·74
189 671	+0·12	170 616	−0·14	182 554	−1·59
188 669	+0·12	170 615	−0·20	182 552	−1·38
188 668	+0·09	170 613	−0·43	183 549	−1·31
187 666	+0·34	170 611	−0·64	184 548	−1·46
182 664	0·00	170 609	−0·64	184 546	−1·53
181 663	−0·21	170 607	−0·61	185 543	−1·38
180 662	0·98	170 605	−0·46	186 542	−1·31
179 660	−0·61	171 603	−0·38	186 540	−1·22
178 658	−0·61	171 600	−0·58	187 539	−1·16
178 657	−0·55	171 599	−0·67	187 537	−1·22
177 655	−0·58	171 597	−0·67	188 535	−1·34
176 654	−0·76	172 594	−0·83	189 533	−1·22
175 652	−0·64	172 592	−0·83	Skipsea	−1·44
174 650	−1·38	172 590	−1·01		
173 649	−1·13	173 589	−1·16	190 530	−1·22
Bridlington	−0·36	173 586	−1·04	190 529	−1·13
		174 583	−1·01	191 528	−0·98
172 646	−0·73	174 581	−1·22	191 527	−1·16
172 644	−0·67	174 579	−1·25	191 526	−1·22
171 641	−0·55	175 578	−1·34	192 525	−1·16
171 639	−0·46	Barmston	−0·52	193 523	−1·22
171 638	−0·46			194 520	−1·22
171 637	−0·43	175 576	−1·53	194 517	−1·22
171 635	−0·40	175 574	−1·46	195 515	−1·10
170 633	−0·18	176 571	−1·56	196 513	−1·16
Carnaby	−0·49	177 568	−1·62	196 512	−1·07
		177 567	−1·68	197 511	−1·04
170 631	−0·12	178 565	−1·59	197 509	−1·07
170 630	−0·06	Ulrome	−1·57	198 507	−1·07

Table 7.1—*continued*

Grid reference	metres p.a.	Grid reference	metres p.a.	Grid reference	metres p.a.
199 505	−1·16	223 449	−1·38	263 390	−1·13
200 503	−1·07	224 447	−1·38	264 389	−1·04
Atwick	−1·13	225 446	−1·31	264 387	−1·34
		226 444	−1·38	266 385	−1·28
200 501	−1·13	227 442	−1·40	268 383	−1·28
201 498	−1·16	228 440	−1·46	269 381	−1·38
202 496	−1·04	229 439	−1·43	271 378	−1·22
203 495	−0·92	230 438	−1·43	273 375	−1·16
203 494	−0·79	231 436	−1·53	275 373	−1·07
204 492	−0·79	232 435	−1·40	276 371	−1·14
205 491	−0·82	233 433	−1·43	277 369	−1·22
206 489	−0·66	234 431	−1·36	Aldbrough	−1·24
206 487	−0·58	235 430	−1·28		
207 486	−0·43	235 429	−1·43	279 366	−1·13
208 484	−0·21	236 428	−1·46	281 363	−1·25
208 483	−0·18	237 426	−1·34	284 360	−1·19
209 482	−0·24	238 425	−1·34	285 357	−1·01
209 480	−0·31	239 422	−1·43	287 355	−1·10
210 477	−0·92	240 421	−1·46	288 353	−1·16
210 476	−0·82	241 419	−1·53	East Garton	−1·14
211 475	−0·82	243 418	−1·46		
211 473	−0·67	244 416	−1·60	290 351	−0·79
212 472	−1·16	245 414	−1·53	292 348	−0·98
213 470	−1·40	246 413	−1·68	294 344	−0·95
213 468	−1·56	248 411	−1·30	296 342	−0·95
214 466	−1·38	249 408	−1·53	298 340	−1·02
215 465	−1·38	Mappleton	−1·45	299 338	−1·10
Hornsea	−0·84			300 337	−1·04
		251 406	−1·38	302 334	−0·92
217 462	−1·31	253 404	−1·38	303 333	−1·04
217 461	−1·22	254 402	−1·22	305 331	−0·92
218 458	−1·53	255 401	−1·34	306 329	−0·85
219 457	−1·46	256 399	−1·62	308 327	−0·76
220 455	−1·53	257 398	−1·13	309 325	−0·92
221 454	−1·62	258 396	−1·16	312 322	−0·85
222 452	−1·68	260 394	−1·22	313 320	−0·79
223 450	−1·65	262 391	−1·16	314 318	−0·73

Table 7.1—*continued*

Grid reference	Metres p.a.	Grid reference	metres p.a.	Grid reference	metres p.a.
315 317	−0·67	349 272	−1·84	382 229	−1·56
316 315	−0·61	349 271	−1·92	383 228	−1·62
318 313	−0·70	350 270	−2·10	384 227	−1·68
320 310	−0·40	351 269	−2·12	386 225	−1·74
Roos	−0·85	353 267	−2·07	387 223	−1·71
		Withernsea	−1·08	388 221	−1·71
321 309	−0·70			389 220	−1·62
322 308	−0·89	354 265	−1·89	390 219	−1·80
323 306	−0·85	355 264	−1·62	391 218	−1·74
325 304	−0·92	357 262	−1·46	391 217	−1·77
326 302	−0·90	359 260	−1·10	392 216	−1·53
327 301	−0·90	361 257	−1·16	394 214	−1·53
329 299	−0·89	362 256	−1·16	395 212	−1·45
331 297	−0·85	363 254	−1·10	396 211	−1·56
332 295	−0·76	365 252	−1·22	397 209	−1·57
Rimswell	−0·85	366 250	−1·45	398 208	−1·68
		368 248	−1·43	399 206	−1·83
334 292	−0·79	Hollym	−1·36	400 204	−1·90
336 290	−0·64			401 203	−1·86
337 288	−0·56	369 246	−1·40	402 201	−1·89
339 286	−0·61	371 243	−1·56	403 200	−1·76
340 284	−0·43	373 241	−1·43	404 198	−1·69
342 282	−0·40	374 239	−1·46	404 197	−1·77
343 281	−0·12	376 237	−1·53	405 195	−1·77
344 279	−0·20	378 235	−1·53	406 194	−1·83
345 278	−0·82	379 233	−1·50	406 193	−1·71
346 276	−1·07	380 232	−1·59	407 192	−1·80
347 275	−0·98	Holmpton	−1·50	Easington	−1·89
348 273	−1·77	381 230	−1·71		

the fact that the 1:25,000 and 1:10,560 maps use a contour interval of 25 ft (7·6 m.). The estimates that have been obtained are shown in the table opposite.

The height of the cliffs above O.D. was found in the following manner (Valentin, 1954c). Using a clinometer, the angle of the slope of the cliff was measured and the length of the slope obtained with the use of a 100-ft tape. Given these two observations, it is easy to

TABLE 7.2
Land Loss by Parish, 1852–1952

No.	Parish	Annual cliff recession (m.)	Shore length (m.)	Annual land loss (sq. m.)	Average cliff height (m.)	Annual loss in volume (cu. m.)
1	Bridlington[1]	0·36	4,875	1,755	13·5	23,693
2	Carnaby	0·49	1,350	662	9·9	6,549
3	Barmston	0·52	5,700	2,964	7·0	20,748
4	Ulrome	1·57	1,175	1,845	7·1	13,098
5	Skipsea	1·44	3,600	5,184	10·8	55,987
6	Atwick	1·13	3,075	3,475	16·6	57,681
7	Hornsea	0·84	4,175	3,507	12·6	44,188
8	Mappleton	1·45	6,600	9,570	17·4	166,518
9	Aldbrough	1·24	4,600	5,704	18·0	102,672
10	East Garton	1·14	1,925	2,195	22·6	49,596
11	Roos	0·85	5,275	4,484	17·0	76,224
12	Rimswell	0·85	1,950	1,658	13·4	22,211
13	Withernsea	1·08	3,600	3,888	10·6	41,213
14	Hollym	1·36	2,250	3,060	11·9	36,414
15	Holmpton	1·50	2,000	3,000	15·3	45,900
16	Easington[2]	1·89	9,375	17,719	13·6	240,975

[1] From Sewerby.   [2] East coast up to Kilnsea Warren.

calculate the height of the cliff crest above its base. To this must be added the difference in altitude between the foot of the cliff and O.D., which was estimated to be 3 m. It is important to work from O.D. rather than from the foot of the cliff, since erosion is not confined to land above high-tide level. The approximate figures for the volume of land lost are shown in Table 7.2.

Although it is important to know the rate of land loss in administrative districts, it is much more useful to consider the recession in relation to the nature of the coast. The stretch of coast under study may be divided into six sections, beginning with Flamborough Head in the north. This is followed southwards by four sections of cliffed coastline:

1. Sewerby to Earl's Dike.
2. Earl's Dike to Hornsea.
3. Hornsea to Withernsea.
4. Withernsea to Kilnsea Warren.

Finally, Spurn Head completes the stretch to the south. No precise measurements have been taken on the first section of coast, i.e. Flamborough Head, but there are some earlier data that we may note. Sheppard (1912, p. 4) quoted E. R. Mathews's figure for a recession of 1·83 m. per annum, but this is undoubtedly an overestimate. As Steers (1948, p. 409) pointed out, the chalk in this vicinity erodes slowly, and Minikin (1952, p. 8) judged the annual retreat to be 0·30 m. Furthermore, the cliffs of glacial material immediately to the south-west retreated only 0·18–0·24 m. annually during the hundred years from 1852 to 1952, and it seems most unlikely that the harder chalk could have receded faster than 0·15 m. each year.

Sewerby to Earl's Dike

This is the most northerly cliff section of Holderness, extending for 8·1 km.; the cliffs are composed of glacial loam capped by loose boulders. The rate of erosion is generally less than elsewhere, although highly variable from one spot to another. Some land has been reclaimed from the sea by the construction of a promenade north of the port of Bridlington, whereas the promenade farther south was built only after a considerable amount of erosion had taken place. The town's 3·5-km. frontage is now stabilised by sea walls and groynes. At the southern end of the seafront, at point 174 650, there is an interesting example of erosion having been replaced by accretion. The hundred-year average rate of recession at this point is 1·83 m., and erosion continued after the promenade was built; in more recent times, however, some small dunes have developed at the foot of the cliff, which is now heavily vegetated. The opposite sequence of events has occurred 3 km. farther south, at point 170 620. The dunes at Auburn that had collected at the foot of the cliff since 1852 have recently been destroyed, with the result that a row of concrete blocks set along the foot of the dunes in 1941 now stands on the foreshore, and the bunkers, etc., that were built on the crest of the cliff now lie on the beach. The clay beds are being attacked and the cliff is still retreating.

In view of these changes from erosion to accumulation and vice versa, data were collected for the period 1941–52 for the 3·7 km. of coast between Bridlington and Earl's Dike. Evidence was obtained from the positions of the virtually immovable concrete blocks in relation to the cliff foot, the situation of the bunkers, the appearance of the cliff and information from local people. The results of these

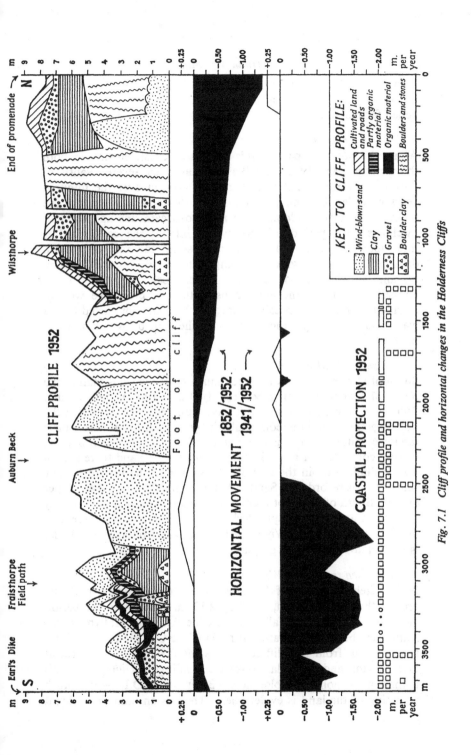

Fig. 7.1 Cliff profile and horizontal changes in the Holderness Cliffs

inquiries are shown in Fig. 7.1, in which it is evident that the concrete blocks have inhibited erosion by comparison with what has happened farther south.

Earl's Dike to Hornsea

Along the northern part of this section of coast, erosion in both the periods 1852–1952 and 1941–52 has been severe. South of Ulrome and Skipsea, the mean rate of erosion over the hundred years exceeded 1·50 m. per annum. In this area one can see the remnants of farm buildings; the numerous caravans and summer dwellings must periodically be moved inland; the coast road is interrupted at point 184 547. Farther south the rate of erosion declines, and near Atwick the average rate of loss was about 1·0 m. per annum, while at Hornsea the rate was locally as low as 0·18 m. annually (owing to the early construction of protective works). As at Bridlington, Hornsea's frontage of 1·5 km. is protected by sea walls and groynes (cf. the diagram of the wall by Minikin, 1952, p. 20).

Hornsea to Withernsea

South of Hornsea erosion has been much more vigorous than to the north of that town, being between 1·40 and 1·68 m. per annum. Sections of the boulder-clay cliff have slumped, carrying down the buildings that stood thereon, while the grass sward is almost intact; the lower cliff so formed is being attacked by the sea. The rapid loss of land continues for a few kilometres southwards and there is then a slow diminution in the rate of recession until the first really low rate of loss is recorded at Sand le Mere (point 320 310) – 0·40 m. per annum. South of this spot there was a slight increase in the severity of erosion and then a rapid decline to the lowest measurement of 0·12 m. per annum at Withernsea, where the frontage of 1·3 km. is protected by walls and groynes.

Withernsea to Kilnsea Warren

Immediately south-east of Withernsea the rate of erosion is markedly greater than to the north, reaching 2·12 m. annually in the period 1852 to 1952. The rate falls again in the parish of Hollym to the south-east, but rises thereafter to 1·50 m. The Cliff House at Holmpton is 80 m. from the cliff top, the same distance as the farm at Out Newton and the radar tower on Dimlington High Land; this is the highest point in Holderness – 38 m. above O.D. – and much slumping of the land is evident here. There is an immediate threat to Dimlington Farm at point 398 208, where the mean annual recession was 1·68 m. although the partial cover of vegetation on the cliff in 1952 was indicative of a slower rate of erosion. In a case such as this, the value of photographs to document changes in the landscape is clearly demonstrated. In the extreme south-east of Holderness, between Easington and Kilnsea, the rate of erosion was catastrophic; for more than 3·5 km. of coast the mean annual recession during the hundred-year period exceeded 2·0 m., and in two places the rate was as high as 2·75 m. This extreme figure is among the highest to be recorded over a comparable period of time anywhere in the world. The glacial upland of Holderness here becomes broken into isolated heights, connected by sand dunes along the east coast and by marsh along the south (Humber) shore. In normal years the cliffs and dunes recede at the same pace, but in exceptional storms, as in 1905–6 and 1953, the line of dunes has been breached by the sea and the whole marsh area flooded from sea to river near Skeffling (Sheppard, 1912, p. 110; Steers, 1953b, p. 286).

Summary on the rate of erosion

TABLE 7.3

Land Loss by Natural Sections of Coast, 1852–1952

No.	Section	Annual cliff recession (m)	Shore length (m.)	Annual land loss (sq.m.)	Ave. cliff height (m.)	Annual loss in volume (cu.m.)
A	Sewerby–Earl's Dike	0·29	8,100	2,357	11·0	25,927
B	Earl's Dike–Hornsea	1·10	13,650	15,015	11·8	177,177
C	Hornsea–Withernsea	1·12	24,250	27,160	16·2	439,992
D	Withernsea–Kilnsea Warren	1·75	15,525	27,200	13·2	359,040
	Entire Coast (approx.)	1·20	61,500	72,000	14·0	1 million

The loss of land in the four sections of coast may be summarised as follows. Between Sewerby and Earl's Dike the mean loss was small, but the rate varied considerably; from Earl's Dike there was a rapid increase in the tempo of erosion and then a slow decline south-eastwards towards Hornsea, a pattern repeated between Hornsea and Withernsea; finally, between Withernsea and Kilnsea Warren erosion increased rapidly in a south-easterly direction, then decreased slightly, culminating in the extreme rates between Easington and Kilnsea. One may imagine a wave-like pattern of successive high and low rates of erosion along the coast, with especially active recession in the extreme south-east. This pattern is shown in Table 7.3 and more clearly in Table 7.2; the former table shows that the mean recession for the whole coast has been 1·20 m. annually between 1852 and 1952. Altogether approximately 720 hectares of land have been lost, or 100 million c.m. of rock, which, having a specific gravity of 2·10, is equivalent to 210 million metric tons of material.

The question arises as to where this large mass of material has been deposited. Flamborough Head and Spurn Head are both outside the area of this study, but a brief look at the development of the latter is worth while. Owing to the lack of accurately surveyed reference points, it is impossible to carry out careful measurements with a tape in the manner employed in Holderness. Consequently it is necessary to compare the oldest and the newest 6-in. maps, superimposing the one on the other (Fig. 7.2). Along the north-east section of Spurn one can note the gradual decline in the rate of loss towards the south, while the westerly movement of dunes on the north-west part results in a compensating gain in land area. Farther south the westerly movement of the coastlines is negligible. However, what were islands in 1852 have been joined together so that the spit has become both wider and longer. The net effect is a gain of just under 50 hectares of land on Spurn Head, which compares with the loss of 720 hectares in Holderness to the north. When the low elevation of Spurn Head is considered – the dunes rise to 10 m. above O.D. and the mean height of the spit is certainly less than 6 m. O.D. – the discrepancy is even greater. Let us suppose that all the land gained on Spurn Head between 1852 and 1952 has been built up from mean sea-level to 6 m.; there is then an accretion of only 3·0 million cu. m. of material to offset the loss of 100 million cu. m. farther north. The remaining 97 per cent af the eroded material must have been deposited in the North Sea, in the Humber or farther south along the Lincolnshire coast.

However, the main purpose of this inquiry is to determine the causes of the loss of land in Holderness, not to trace whither the material has been transported, interesting though that question is.

II. REASONS FOR THE LOSS OF LAND

To explain everything simultaneously is impossible and therefore one must approach the problem by stages:

    1. The general causes of cliff recession.

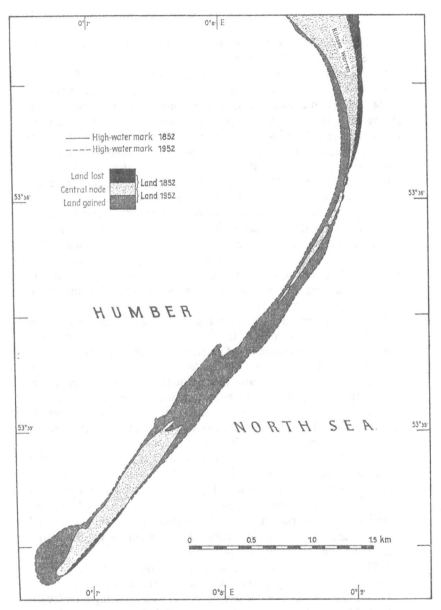

Fig. 7.2 High-water marks at Spurn Head, 1852 and 1952

2. The reasons for the increase in the rate of erosion towards the south-east.

3. The reasons for the alternations between rapid and slow recession along the coast.

It is not necessary to review all the general causes of cliff erosion, as this has been done repeatedly in the last few years. Steers (1953*a*, p. 24) has enumerated the various sub-aerial factors relevant for Holderness, including the seepage of ground water, the effects of which the present author has also observed. The single most important factor in cliff recession is undoubtedly the action of the waves, which are most effective during high water in storms. The storm surge of 31 January 1953 produced remarkable changes in the coast of Holderness. Lack of time precluded accurate measurement by the author of this recession at all the 307 points along the coast that had previously been determined (see above), but measurements at sample points showed that the retreat of the shore was some multiple of the annual rate over the hundred-year period. W. W. Williams described even more severe damage near Lowestoft in Suffolk, where he observed a cliff recession of between 9 and 26 m. in a period of seventeen hours.

At ordinary spring high tides the waves reach the foot of the cliffs along the greater part of the Holderness coast. During medium high tides only some stretches of cliff are attacked by the waves, as at Dimlington High Land. Neap tides leave all but a few sections of cliff unaffected. At other states of the tide the sea does not reach the foot of the cliff. The obvious relationship between the level of the water and the rate of erosion suggests that differences in the observed rate of erosion along the coast may be due to variations in the relative height of the water surface. These variations have been analysed by Valentin (1954*b*, p. 21), using British tide-gauge records. The evidence for this proposition should be clearest in those sections of coast where the rate of erosion has been greatest. For the thirty points on the Holderness coast for which the observed average recession over the period 1852–1952 exceeded 1·50 m., additional measurements were made from the O.S. 6-in. maps to obtain the mean recession for the following periods:

1852–1889: 1·53 m. per annum (a time of low sea-level).

1889–1908: 1·77 m. per annum (a time of rising sea-level).

1908–1926: 1·89 m. per annum (a time of averagely high sea-level).

1926–1952: 1·71 m. per annum (a time of slightly falling sea-level).

The average figures shown above conceal much variation among the thirty points of observation. However, the recorded rates of erosion are consistent with the proposition that there is a relationship with the height of the sea's surface; erosion has been faster when the sea-level has been high than when it has been low. (For the changes in sea-level, see Valentin, 1954*b*, Pt. 2.)

Changes in sea-level could be one of the causes for the greater rate of recession in the south-east. As a result of the isostatic readjustment of the land in Post-glacial times, the southern part of Holderness sinks about 15 mm. per century relative to the northern part (Valentin, 1953*a*, 1954*b*, Pt. 1). However, this is sufficient to play only a very small part in the explanation of the north-south difference in the tempo of erosion, and one must look for other explanations, especially in the exposure of the coast to wave attack.

Bridlington Bay is sheltered by Flamborough Head, which juts out into the sea for 8 km., and by the Smithic sandbank which lies offshore; this sandbank is 10 km. long and at spring low water rises 2·7 m. out of the sea. Even during stormy weather only medium-sized waves reach the cliffs which back Bridlington Bay, which accounts for the slow recession of the land despite the fact that the cliffs are comparatively low and Flamborough Head itself acts as a large groyne to prevent the coastwise transport of material from farther north. By contrast, the middle section of the coast has little shelter and the southern part bears the full brunt of northerly, north-easterly and easterly storms, as well as the swell from the same directions. Furthermore, in the south the sea bed shelves steeply; at Dimlington High Land, for example, the 10-m. submarine contour is only 600 m. from the beach. As a consequence the full impact of the waves is felt at the foot of the cliff. Hence, despite the fact that the cliffs are high, and despite the transport of material from the north, the land recedes rapidly.

It is useful to summarise the findings cartographically, as in Fig. 7.3. Landward from the present-day coast is a profile (shaded in grey) showing the absolute height of the cliff above sea-level, averaged over the hundred years, and exaggerated 300 times compared with the scale of the map. This profile has been compiled from the data afforded by the 307 points, the height of the cliff being drawn perpendicular to the line of the modern coast. This profile could be improved by including geological data, but this has not been done since the glacial material is comparatively uniform. A similar profile has been drawn to show the accumulated loss in land over the same period of time, and also the localised

ACG E

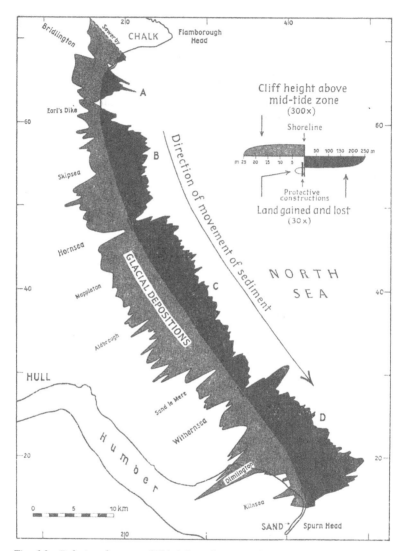

Fig. 6.3   Relations between cliff height and erosion along the coast of Holderness

gains; the former are shown seaward of the present coast and the latter landward. The scale of this profile is exaggerated thirty times, so that the map shows the position of the coast as it would have been 3,000 years ago if the rate of erosion over the whole period since then was the same as that between 1852 and 1952.

It is unlikely that the rate of recession has been constant over the last 3,000 years, since we know that there have been several changes of sea-level since 1000 B.C., in particular a rise of sea-level relative to the land. The rate of erosion has therefore probably increased in recent times. Moreover, as is shown below, the irregularity of the recession in the sections of coast about Bridlington, Hornsea and Withernsea is a modern phenomenon. Nevertheless, for many hundreds of years there has been faster erosion in the southern part of the coast than in the northern section, partly on account of varying degrees of exposure. Consequently the real coast of 3,000 years ago was a little farther west than is shown in the shaded area in Fig. 7.3 and was smoother than indicated. In general, there has been little erosion in Bridlington Bay and a noticeable increase to the south-east, until at the extremity of the section the retreat may be greater, probably as much as 7 km.

This interpretation does not agree with that of Sheppard, who considered that the Roman coastline between Flamborough Head and Spurn Head lay approximately 2·5–3·5 km. east of the present coast (Sheppard, 1912, p. iv; cf. Brown, 1930, p. 315; Steers, 1948, p. 410; Sölch, 1951, p. 503). The interpretation put forward in the present paper is supported by recent geomorphological work undertaken by the author (Valentin, 1953b, 1954c). During the retreat of the last ice, when the front of the ice sheet ran from Flamborough Head across the Smithic Bank to Ulrome, a basin in the boulder clay was exposed in the northern section of the coast, in which clay varves were deposited. Post-glacial deposits were laid over the varves, as at Fraisthorpe and Barmston, and among the fresh-water shells have also been found marine shells above the highest level reached by the sea at the present time. This implies that, probably around 2000 B.C., the North Sea extended much farther west than it does today.

By contrast, in the middle section of the coast morainic deposits must have been exposed as land farther east than is now the case. Evidence for this proposition is found in the east–west and north-east/south-west trending subglacial channels that converge on Hornsea Mere, in which only fresh-water deposits were laid down;

similar materials were formed in one-time lakes to the east of the Mere and also in basins within Aldbrough channel. Among the sandy hills of Sand le Mere and Withernsea, south-east of Aldbrough, are deposits which hand-drilling has shown to attain a depth of 13 m. and which have characteristics indicative of predominantly brackish water. This brackish water did not originate from the north-east, from the North Sea, but came instead from the south-west and the Humber. Confirmation of these facts is found in the remnants of a terminal moraine along the present coast in the vicinity of Atwick, Mappleton, Grimston and elsewhere. To judge by the predominantly south-westerly trend of the drainage and the few spot-heights shown on the 1852 edition of the 6-in. map, this moraine must have reached a considerable height, exceeding that of the terminal moraine which runs by Bewholme, Rise, Sproatley, Keyingham and Patrington farther to the west.

Near Holmpton and Dimlington, in south-east Holderness, the two terminal moraines are close together and run parallel in an easterly direction into the North Sea. A hundred years ago Dimlington High Land extended 175 m. farther east and rose to a height of 42 m. There can be no doubt that a specially high ridge of terminal moraine extended seawards, and at the end of the last glacial period may have reached New Sand Hole. This depression can be interpreted as one of a series of radial channels formed under a lobe of ice which extended from Yorkshire to the Dogger Bank. The channels include those already mentioned in Holderness and the Silver Pit, Outer Dowsing, Sole Pit, Coal Pit, Well Hole, Markham's Hole, Outer Silver Pit and Skate Hole, etc. At the start of the last glacial period the Dogger Bank formed a pronounced ridge. It represents a continuation of the Chalk and Jurassic series which outcrop in Yorkshire and also in Denmark and Scania. On the Dogger Bank these rocks are capped by deposits of an early glaciation.

After the last glaciation, therefore eustatic movements caused the North Sea to penetrate in a southerly direction into the basin between Yorkshire and the Dogger Bank. It would have reached the Chalk of Flamborough Head at a point a little east of the present headland, and immediately to the south of this cape entered the low plain composed of varve deposits, thus almost reaching the line of the present coast. Farther south the advancing sea was checked by the moraines of Holderness, especially the high terminal ridge of Dimlington that extended in a south-easterly direction. The main attack of the waves was concentrated against this ridge.

Hence, rather as the hands of a clock rotate about their axis, the coast began to retreat on the fulcrum of Flamborough Head. Even after several thousand years the high land of Dimlington still forms a marked feature in the coast. The present S-shaped alignment of the coast between Bridlington Bay and Spurn Head is largely, therefore, the result of the relief as it existed at the end of glacial times; this alignment cannot be explained simply by the present coastal processes, which erode the land rapidly in the south-east but more slowly in the north-west. As erosion destroys Dimlington High Land, so must the whole coast between Bridlington Bay and Lincolnshire retreat.

III. REASONS FOR THE ALTERNATION BETWEEN RAPID AND SLOW EROSION ALONG THE COAST

Fig. 7.3 shows that north and south of Hornsea there is some relationship between the profiles of cliff height and cliff recession, which suggests that the slower recession south-east from Hornsea is related to the greater cliff height. This explanation would be natural if one could assume equality of geological conditions and of exposure to the sea's attack along the section of coast. However, exposure increases southwards, and since this factor appears to be more important than cliff height (see p. 129), it is doubtful if the two profiles are in fact closely related. Elsewhere along the coast, especially at Bridlington, Hornsea itself and around Withernsea, the two curves are as mirror-images of each other, proof that some other factor has been operative.

The coast-protection works built after 1852 have had an important influence on the rates of coast recession. The effect of the sea walls in reducing land loss is localised to the stretch of coast actually protected. On the other hand groynes have an effect in checking the southward movement of beach material which begins several kilometres northward of the groynes, and also checks the rate of cliff recession. This reduction is compensated for by increased erosion south of the groynes. Evidence for this is seen in the Bridlington section of coast, where the recession was very small north of the harbour jetty, and some land was even gained at the time the promenade was built. When this promenade and the associated groyne system was extended farther south, the retreat of the protected cliffs ceased altogether; by contrast, the rate accelerated at the south end of the protective works (cf. p. 122). This influence of groynes is also clear in the middle section of the coast, as at Hornsea and Sand le Mere, but most markedly at Withernsea. However, the coast-protection works built in 1914 at Kilnsea have so far had only a small effect on the rate of coast recession.

IV. THE STRUGGLE AGAINST COAST EROSION

The present loss of 7·2 hectares of land per annum suggests that the gaps between the existing coast-protection works should be closed. There is no doubt that this is technically possible. However, the cost would be enormous compared with the value of the land saved. The Town Clerk of Bridlington, who was also secretary of the East Yorkshire Joint Advisory Committee for Coast Protection Works, kindly made available several unpublished engineering reports that were prepared in 1949–50, in which the cost per mile of sea wall plus groynes was estimated at that time at £310,000. To protect the whole of the 55 km. of unguarded coast would cost £10·6 million. The capital value of the land which is now lost each year is £1,040, which means that the protective works would have to be effective for at least 10,000 years to recoup the outlay. The complete cessation of cliff erosion along the whole of the Holderness coast would eliminate the main source of beach material, with the consequence that the waves would slowly remove the foreshore and so expose the foundations of the protective works. Thus the walls and groynes would inevitably be destroyed, and it is impossible to envisage them lasting 10,000 years. In addition to the cost of construction, allow-ance must be made for annual repairs and maintenance, which may be estimated at 4 per cent of the capital cost, i.e. £424,000, which is 400 times greater than the value of the land lost each year.

If complete protection is out of the question, let us look at the possibility of defending the critical points upon which the fate of the coast and of the hinterland depends. For this purpose, reference should be made to Fig. 7.3 and to the 1:100,000 topographical map of Holderness (in Valentin, 1954c). Owing to its protected situation, the north may be regarded as safe for the time being, though a rising sea-level presents the danger that salt water may break into the low valleys of Earl's Dike and especially the Barmston Main Drain. The first critical point to note is at Hornsea, where the sea may penetrate along the subglacial channel to Hornsea Mere. It would not be possible for the sea to reach much farther west on account of a rise in the valley floor near Seaton. The position is worse in the more southerly valley of Sand le Mere, where a breach by the sea would enable marine waters to reach the Humber via the south-west-trending drainage system that passes by Roos and Keyingham; large tracts of lowland would be flooded and south-east Holderness would be cut off from the mainland. The same could happen in the valley of Withernsea; levels run by the author showed that near Patrington the land is 1·80 m. above O.D., which means that it is 1·50 m. below water-level at a medium spring high tide and about 2·60 m. below the highest tide level of the year. However, in Withernsea itself the land is a little higher and is in any case already protected. Thus it is quite clear that these three points – Hornsea, Sand le Mere and Withernsea – must be protected now and in the future by all available means.

The much greater importance for coast protection of Dimlington High Land appears hitherto not to have been recognised. Dimlington High Land represents the last remains of a high ridge of terminal moraine that stretched far to the south-east. This ridge has checked the rate of coast recession, but as it has been destroyed conditions have become worse both to the north and to the south. In 1852 Dimlington High Land was 42 m. above O.D., but it now stands only 38 m. high, and in another century it is likely that the cliff there will be only 30 m. above O.D., with the prospect of a further rapid decline in elevation thereafter as the coast continues to retreat. The immediate effect of the reduction in height will be an acceleration of the pace of erosion. Already there is a serious problem south of Dimlington, at Easington and Kilnsea, where the isolated glacial uplands are being eroded with extraordinary rapidity; the sea has twice broken through the low dune and marsh land between the two settlements. Spurn Head itself, which begins at this point, is in any case only held in place with great difficulty for its entire length by groynes and sea walls. If Spurn Head were to disappear, this would make further reclamation planned on the Humber flats impossible, as well as changing the hydrographical conditions of the commercially important river channel, which in turn would affect industrial and agricultural lands now protected by dykes. The implications of such changes cannot be estimated. Even the coast of Lincolnshire is threatened, as witness the damage caused by the storm of 31 January 1953. Dimlington High Land is the last bastion against the sea for the whole stretch of coast between Flamborough Head and east Lincolnshire, and continued recession at Dimlington will have widespread repercussions that could well cost millions of pounds. Fortunately it is not too late to prevent, or at least mitigate, these dangers, and coast-protection works from Dimlington High Land to Spurn Head must therefore be regarded as of great national urgency.

Even if the critical sections of coast were protected, the sea would still attack 46 km. of cliff that remain exposed. This might seem advantageous, since the supply of beach materials would be maintained, something that is especially necessary to maintain the defences of the south-east section of the coast. Undoubtedly a wide and high beach is the best possible shield against the sea. But a few centuries hence the situation would be less satisfactory, because the protected sections would jut out into the sea as artificial promontones. Defence would become continually more difficult on the sea fronts, while on the flanks of the promontories new protective works would become necessary. The unprotected stretches of coast would retreat to form wide bays, but there is little hope that the power of the waves would decline and that ultimately the cliffs in the bays would cease to recede. There can also be scant expectation of any favourable change in the natural conditions, such as a fall in sea-level. Nevertheless, the rapid development of knowledge in the past century allows us to be optimistic that in the future some economic means may be found for checking the erosion of Holderness, and we need not therefore despair.

SUMMARY

Along the 61·5 km. of the Holderness coast, 307 points were selected at which to measure the retreat of the land between 1852 and 1952. On average, the cliffs receded by 120 m. in that period, approximately 720 hectares of land were lost and hence about 100 million cu. m. of material were carried away, of which no more than 3 per cent appears to have been incorporated in Spurn Head. In general, erosion has been less severe in the north than in the south-east; furthermore, a wave-like alternation of fast and slow erosion has been evident along the coast. Four natural sections of coastline can be recognised.

The most important general cause of erosion is the energy of the sea, and a connection is shown to exist between the rate of recession and the level of the water's surface. The faster loss of land to the south-east is related to the greater exposure to wave attack. Dimlington High Land, the last remnant of a terminal moraine that formerly extended farther south-east, forms high ground that is attacked vigorously by the waves. The alternations between fast and slow erosion are primarily caused by coast-protection works. To extend the protective works to the whole coast would be prohibitively expensive with present-day techniques. It is therefore necessary to select the key points on the coast where defences should be built: these are Hornsea, Sand le Mere, Withernsea and especially south-east Holderness from Dimlington to Spurn Head. For the long-term future one must hope that technical development will yield more economical methods of coast protection that will enable the loss of land in Holderness to be checked. Plates 4 and 5 show Spurn Head and Holderness respectively.

REFERENCES

BROWN, R. N. R. (1930) 'Holderness and the Humber', in A. G. OGILVIE (ed.), *Great Britain: Essays in Regional Geography*, 2nd ed. (Cambridge), pp. 312–21.

MINIKIN, R. R. (1952) *Coast P Erosion androtection: Studies in Causes and Remedies.*

REID, C. (1885) 'The geology of Holderness, and the adjoining parts of Yorkshire and Lincolnshire', *Mem. Geol. Surv.*

SHEPPARD, T. (1906) 'List of papers, maps, etc., relating to the erosion of the Holderness coast, and to changes in the Humber estuary', *Trans. Hull Geol. Soc.*, vi (1) 43–57.

—— (1909) 'Changes on the east coast of England within the historical period: I. Yorkshire', *Geogr. J.*, xxxiv (5) 500–13.

—— (1912) *The Lost Towns of the Yorkshire Coast and Other Chapters Bearing upon the Geography of the District.*

SÖLCH, J. (1951) *Die Landschaften der Britischen Inseln: I. Bd., England und Wales* (Vienna).

STEERS, J. A. (1953a) *The sea coast.*

—— (1953b) 'The east coast floods, Jan 31–Feb 1, 1953', *Geogr. J.*, cxix (3) 280–98.

—— (1969) *The Coastline of England and Wales*, 2nd edn (Cambridge).

THOMPSON, C. (1923) 'The erosion of the Holderness coast', *Proc. Yorks. Geol. Soc.*, xx (1) 32–9.

VALENTIN, H. (1952) 'Die Küsten der Erde: Beiträge zue allgemeinen und regionalen Küstenmorphologie', *Pet. Mitt.*, Ern.-H., ccxlvi.

—— (1953a) 'Present vertical movements of the British Isles', *Geogr. J.*, cxix (3) 299–305.

—— (1953b) 'Young morainic topography in Holderness', *Nature*, clxxii 919–20.

—— (1954a) 'Gegenwärtige Niveauveränderungen im Nordseeraum', *Pet. Mitt.*, xcviii (2) 103–8.

—— (1954b) 'Gegenwärtige Vertikalbewegungen der Britischen Inseln und des Meeresspiegels', *Verh. 29. Dtsch. Geographentags* (Essen, 1953).

—— (1954c) 'Glazialmorphologische Untersuchungen in Ostengland: Ein Beitrag zum Problem der letzten Vereisung im Nordseeraum' (in preparation).

8 The Coastal Landslips of South-East Devon

(*With an Appendix written specially for this volume*)

MURIEL A. ARBER

I. INTRODUCTION

CHRISTMAS 1939 marks the centenary of the most spectacular landslip recorded on the English coast. The great chasm then formed at Dowlands (Plate 8) and Bindon, near Axmouth, in south-east Devon, was one of a series of slips which have taken place during historic times in a continuous belt along the coast from the mouth of the river Axe on the west to the county boundary between Dorset and Devon on the east. Another related landslip lies farther to the west, at Hooken Cliff between Branscombe and Beer Head. The whole area is included in the Geological Survey (New Series) 1-in. map, Sheets 326 and 340.

All these falls are associated with the Cretaceous overstep of Chalk, Upper Greensand and Gault on an eroded surface of Triassic, Rhaetic and Liassic clays. The general succession may be summarised as follows:

Clay-with-Flints.

Chalk { Middle.
{ Lower.

Upper { Calcereous sandstones and chert beds.
Greensand { Foxmould (sands).
{ Cowstones (lenticular concretions of calcareous
{   sandstone).

Gault (sandy clay).

Lower (Blue) Lias (limestones and shales).

Rhaetic (limestones, shales and marls).

Keuper Marls.

The sub-Cretaceous plane of denudation, which truncates in turn the Keuper Marls, Rhaetic and Blue Lias (see Fig. 8.1), dips slightly towards the sea, the overlying Cretaceous beds having a similar inclination, of about 5°. Water, after percolating through the Chalk and Upper Greensand, accumulates in the Foxmould, over the impervious foundation formed by the argillaceous Keuper, Rhaetic, Lias and Gault. The supporting sands are thus liable to be washed away from the base of the Upper Greensand, so that the chert beds and calcareous sandstones may then founder and slide down towards the sea, carrying the Chalk with them.

II. HOOKEN CLIFF

At Hooken Cliff, below South Down Common, where the plane of unconformity between the Triassic and the Cretaceous beds dips below sea-level to the east (see Fig. 8.1), the Keuper Marls are overlain directly by about 50 ft of Foxmould (Upper Greensand), the

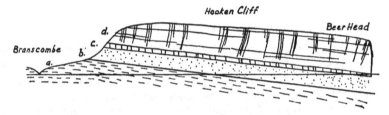

Fig. 8.1   *Coast section from Branscombe to Beer Head*
Horizontal scale: 3·5 in. to 1 mile.
Vertical scale: 1 in. to 900 ft.
(*a*) Keuper Marls.   (*b*) Upper Greensand.
(*c*) Lower Chalk.   (*d*) Middle Chalk.
Based by permission on Strahan and Jukes-Browne, in A. J. Jukes-Browne, *Cretaceous Rocks of Britain*, I, fig. 64 (1900).

Gault being absent west of Seaton. Above the Foxmould lie 150 ft of cherty sandstones and from 200 to 300 ft of Lower and Middle Chalk. The direct contact at sea-level between the sandy Cretaceous beds and the Triassic clays forms a plane of weakness along which a landslip occurred in 1790 (Conybeare *et al.*, 1840).

Two years before the fall, a plentiful stream of water, which had hitherto emerged about half-way up the cliff, ceased to flow. It seems likely that this water, by being deflected through the incipient cracks into the sandy beds, had helped to cause the catastrophe.

Until the slip occurred, South Down Common was bounded by a sheer cliff dropping into the sea. During the winter of 1789–90 a fissure opened on the cliff top, and one night, in March 1790, an area of 7 to 10 acres which had been thus cut off suddenly moved seawards, subsiding vertically between 200 and 260 ft. This isolated mass was not completely shattered, but was broken up into columns and pinnacles still bearing relics of cultivation. The land encroached on the sea for a distance of more than 200 yds from the previous coastline, and at the same time the seabottom was forced up into an offshore reef. Fishermen, who the night before had sailed 8–10 ft above their crab-pots, found them in the morning stranded 15 ft above the surface of the water.

An excellent photograph of Hooken Cliff, the undercliff (Under Hooken), and the Pinnacles (Rowe, 1903, Plate X), shows the condition of the landslip rather more than a hundred years after it had occurred. The extent of the subsidence is demonstrated by Rowe's diagrams, which mark the corresponding positions of zones and individual beds in the cliff and in the isolated pinnacles.

III. THE AXMOUTH LANDSLIPS

The collective term 'Axmouth Landslips' is a convenient designation for the series of cliffs and undercliffs extending from the mouth of the river Axe by Culverhole Point to the county boundary at Devonshire Head, near Lyme Regis, since the greater part of this area lies in the parish of Axmouth. From Culverhole Point eatswards for approximately 4 miles, the undercliffs form a tumbled tract about a quarter of a mile wide between the sea and the inland cliffs, which in places rise to 500 ft at the margin of the south-east Devon peneplain. Pinnacles and masses of displaced rock occur in the undercliffs, and at Dowlands and Bindon a chasm separates a great block of isolated land from the main cliff.

The Axmouth Landslips east of Culverhole Point may be subdivided into those of the Bindon, Dowlands, Rousdon, Charton, Whitelands, Pinhay and Ware Cliffs (see Fig. 8.2). The line of inland cliffs is broken at intervals by deep ravines or 'goyles', opening into the undercliff which is almost entirely overgrown. The right of way along the footpath from Seaton to Lyme was the subject of a lawsuit which lasted from 1841 to 1843, when Ames, the owner of Pinhay cliffs, had built a wall across the path to obstruct public access. He was prosecuted in the Queen's name by Joseph Hayward, of Lyme, who eventually won his case (Wanklyn, 1927, p. 253). Ames, although compelled to reopen the way, is said to have been determined that it should not be enjoyed, and resolved to cut off the view by enclosing the path between walls nearly 8 ft high built of Upper Greensand chert blocks. He proceeded with this scheme for more than 100 yds, and the walls have remained ever

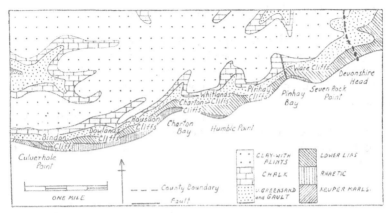

Fig. 8.2 *Sketch-map of coastal geology between Axmouth and Lyme Regis.*
Effects of landslipping omitted.
(Outlines taken by permission from Geological Survey (N.S.) 1-inch map,
Sheets 326 and 340.)

since in the middle of the undercliff. Recently, however, they became
unsafe and had to be partially demolished.

WARE AND PINHAY CLIFFS

It is not known when slipping first began in the cliffs between
Axmouth and Lyme, but the movements are evidently of great
antiquity. While, however, the present forms of the undercliffs
of Bindon, Dowlands and Whitlands have only been attained within
the last century, at Pinhay and Ware, on the other hand, no great
change seems to have taken place over a much longer period.

At Ware the seaward cliffs are formed of Lower Lias, while
those inland consist of Upper Greensand, the Chalk being absent.
The upper surface of the Lias forms a rough tract of ground covered
with Cretaceous debris, and both this undercliff and the main Upper
Greensand cliff are very much overgrown, the only prominent
feature being Chimney Rock, a pinnacle of calcareous sandstone.

At Pinhay, Chalk appears in the inland cliff, in front of which
stands the Chapel Rock (Rowe, 1903, Plate I). Rowe showed that it
is simply the slipped face of the inland cliff, the top of the rock
corresponding in horizon to the top of the main section. According
to tradition, the Chapel Rock derives its name from the secret
worship held behind its shelter during the religious troubles of the

reign of Mary Tudor and in later periods. The subsidence of the Chapel Rock from the main face had thus taken place not later than the sixteenth century, but as recently as 1886 the Great Cleft (Rowe, 1903, Plate II), caused by a slip of similar type a little farther to the west, began to open, and the portion of the cliff thus detached has now sunk into another minor escarpment in front of the main face.

The Pinhay undercliffs are much more distrubed than those at Ware. The densely overgrown hollows, together with the shallower parts which have been planted as orchards, led Jane Austen, in the

Fig. 8.3   Section of Pinhay Cliffs
(a) Chalk.   (b) Greensand.
(c) Lias.   (d) Red Marl.
(Gault and Rhaetic not separately lettered.)
After wood-cut in De la Beche (1830, Pl. 35, Fig. 3)

early nineteenth century, to comment in *Persuasion* on their evident antiquity. Between the sea and the main cliff with its slipped face, the ground lies in a series of three wave-like ridges and gullies running parallel to the coast. At the seaward margin, above the Lias cliffs of Pinhay Bay, the fallen Chalk and Greensand form the tract of Pinhay Warren, but farther to the west the sea cliffs die away, and a chaotic mass of tumbled Chalk reaches down to the edge of the water. De la Beche, in 1830, published a section of Pinhay (see Fig. 8.3) and subsequently reproduced it, with modifications, in his *Geological Manual* (1831 and later editions). His description of the landslipping (1830, p. 61) is remarkable for the fact that although he did not use the term 'Gault', he recognised at this early date that the basal bed of the Cretaceous was a clay, which together with the Lias formed the impervious foundation over which the rest of the Upper Greensand and Chalk had slid.

Dowlands and Bindon Cliffs

From Humble Point to Culverhole Point, inland cliffs of Chalk and Upper Greensand overlook undercliffs of fallen Cretaceous material, thrown up irregularly into mounds and trenches, and sloping down to rest on the solid Lias reefs on the shore. In places a slipped face lies in front of the main cliff, and there are isolated pinnacles of Chalk and Greensand, but the most striking feature is the great chasm at Dowlands and Bindon.

There are various traditions of early falls. On the occasion of the great landslip of 1839 an aged labourer recalled how, many years before, a 'deal of top land sinked down and was spoiled', but the undercliff here was similar to the rest of the series until the Christmas of 1839. Since the previous June the rainfall had been almost continuous, and twice as heavy as usual; the gales also were so long-continued as to wash up on the neighbouring shores large numbers of the tropical *Physalia* or 'Portuguese man-of-war'. A week or two before Christmas, fissures began to open in the cliff top belonging to Dowlands Farm, and on 23 December one of the cottages in the undercliff below, occupied by a labourer named Critchard, began to show signs of subsidence, though not sufficient to create alarm. At one o'clock on the morning of Christmas Day Critchard, with his wife and neighbours, was returning from the old Christmas Eve ceremony of burning the ashen faggot at Dowlands Farm, when he found that the path into the undercliff had begun to sink and that the cottages had subsided still further. About four o'clock there came what Critchard described as a 'wonderful crack', and an hour later he discovered that the door would scarcely open and that the beams of the house were settling, while the garden outside was broken by fissures. He roused the neighbours, and together they cleared their furniture and possessions from the houses. Critchard then scrambled up to the farm for help, but the path had subsided even more, and when a wagon was sent down to fetch their property, the road had to be remade for its return. Meanwhile Critchard's cottage, and another under the same roof, had been upheaved and twisted, while a third cottage, which stood a little apart to the west, was completely destroyed. The movements of the ground took place, however, in absolute silence and without vibration.

This episode proved to be the forerunner of the great landslip which occurred on Christmas night, and was partly seen by the coastguards on duty in the neighbourhood. A strong gale was blowing, but the men, when they went out as usual in the evening, although aware that the cliffs were falling rather more than ordinarily, did not know of the disturbance of the ground on the previous morning. Two coastguards, when walking from Axmouth over the cliff top near Dowlands, stumbled across an unfamiliar ridge of gravel, and soon afterwards one of them jammed his leg in a narrow fissure. By the light of the moon they then saw the land cracking and gaping all round them, and they heard a sound like 'the rending of cloth'. They hastily made their way to safer ground, and proceeded to their station at Whitlands.

Meanwhile, shortly after midnight, two other men who were on duty near Culverhole Point became alarmed by the heaving and disturbance of the beach, the sea being violently agitated while a dark ridge rose in the water. They heard also a deafening crashing of falling rocks, accompanied, as they declared, by 'flashes of fire, and a strong smell of sulphur'.

During 26 December the land that had been cut off by the fissures on the cliff top gradually subsided seawards, and by the evening had reached a position of equilibrium in the undercliff. A new inland cliff, 210 ft high in its central portion and sinking to east and west, had thus been exposed, backing a chasm into which some 20 acres of land had subsided. The length of the chasm was about half a mile, while its breadth increased from 200 ft on the west to 400 ft on the east. The amount of foundered material was estimated at 150 million cu. ft, weighing nearly 8 million tons; most of it was broken into a jumble of small rifted masses and pinnacles of rock, but in places blocks of 2 or 3 acres remained intact though tilted. Beyond the chasm a counterscarp of Chalk bordered an isolated area of 15 acres of cultivated ground, which had moved seawards and subsided to some extent, but on which nevertheless corn- and turnip-fields and hedges survived unshattered.

The sea cliffs of displaced Chalk and Upper Greensand, which had previously stood 50 or 100 ft in height, were now broken and lowered and thrust 50 ft towards the sea, so that the hitherto isolated Pinnacle Rock of Chalk on the shore became relatively inconspicuous. The dark ridge which had been seen rising in the water proved to be a reef of Upper Greensand (cherty sandstone, Foxmould and Cowstones). It stretched for nearly three-quarters of a mile, with its outer edge 300 to 500 ft seaward of the previous high-water mark. The beds were much broken, and now dipped inland at angles varying from 30° to 45°, while the surface, which previously had been at least 10 ft under water at low tide, was now raised in places 40 ft above high-water level, still covered with seaweed, limpets and starfishes. The middle of the reef was joined to the mainland by shingle, but one arm extended freely at the western end, and another to the east enclosed a lagoon which formed a natural harbour. (See Plate 6 and Fig. 8.4.)

By good fortune, two eminent geologists happened to be in the neighbourhood at the time that this great landslip occurred. Conybeare was vicar of Axminster, and Buckland, then President of the Geological Society, was spending Christmas at Lyme. Both visited the scene as soon as possible after the event. The slip was drawn and surveyed at once, and a description by Conybeare, dated

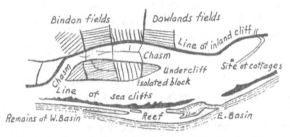

Fig. 8.4 Ground-plan of Dowlands and Bindon landslip
Southward-facing cliffs shown in solid black
Based on wood-cut in Roberts (1840, 5th edition)

31 December 1839, was published in the next issue of Woolmer's *Exeter and Plymouth Gazette*. The early articles and correspondence in the local newspapers and the reports in scientific journals by various observers are not entirely consistent in their accounts of the sequence of events. On the whole the most authoritative statement is that contained in the large memoir (Conybeare *et al.*, 1840) which was published in London by subscription during the following summer. A smaller and more popular guide (Roberts, 1840) written by George Roberts, the historian of Lyme Regis, and published locally by Daniel Dunster, ran into five editions by the end of 1840. Dunster also produced a fine set of lithographic views, mostly by the hand of G. Hawkins, Jr, examples of which are shown in Plates 6 and 7, and a further series was made by L. E. Reed of Tiverton. The landslip was therefore remarkably well recorded while still of recent occurrence.

These contemporary descriptions and illustrations are the more valuable on account of the rapid effect of weathering on the scene.

Buckland's prophecy, made in his original report, that the sea would probably destroy the reef owing to its 'perishable nature' was soon justified. A question is said to have been asked in the House of Commons concerning the possible use by the Government of the eastern basin as a harbour, since it was larger than the Cobb at Lyme. From the first, however, the entrance was too shallow for the use of shipping, and during the following year the pool became choked with debris, while the whole reef was being broken up and removed by the sea.

On the other hand the great chasm even now shows comparatively little alteration from its original condition. The principal change has been the reduction and loss of many of the pinnacles, which are further obscured by the luxuriant growth of vegetation. A photograph by Mr W. E. Howarth (Lang and Thomas, 1936) shows the appearance of the chasm at Whitsuntide in 1936.

Conybeare's memoir of the Dowlands landslip included a consideration of the causes of the subsidence. His account of the effect of springs and rainfall in washing away the Foxmould, and thus undermining the Cretaceous beds, was substantially the same as De la Beche's description of the slipping at Pinhay, but the Dowlands problem was complicated by the existence of the chasm and reef. Conybeare believed that the Foxmould, which was from 150 to 200 ft thick, was probably drained dry in its upper part, while the lower layers were reduced to the condition of a quicksand by water held up over the argillaceous surface of the underlying Lias (from which he did not here distinguish the Gault), and water was further impounded in the undercliff by the fallen rock masses of previous landslips. In this way the upper Foxmould, together with some 200 ft of cherty sandstones and Chalk above, came to founder into the lower Foxmould, slipping somewhat seawards down the dip, and thus forming the chasm and isolating the outer block of cliff, which had only sunk to a minor degree. According to Dawson (Conybeare *et al.*, 1840) the volume and weight of the subsided mass were sufficient to account for the forcing-up of the sea floor into a reef, since the Foxmould was there overlain by only 30 ft of cherty sandstones under shallow water, so that resistance was at a minimum. Conybeare's views were illustrated by a large section across the landslip, and also by a simplified diagram which is here reproduced (Fig. 8.5); his explanation and figures were accepted as satisfactory by the officers of the Geological Survey in 1911.

The general public were not, however, satisfied by the rational views of geologists concerning the cause of this 'most extraordinary and terrific explosion of nature', and despite the local character of the movements and the total absence of vibration, popular opinion attributed the event to the agency of a volcano or an earthquake. The cliffs east of Lyme were already associated with the idea of vulcanicity on account of the spontaneous combustion of pyrites in the Lias, and the reports of the coastguards, who had declared that they had seen flashing lights and smelt a sulphureous smell, were considered to confirm the volcanic theory, although it is now

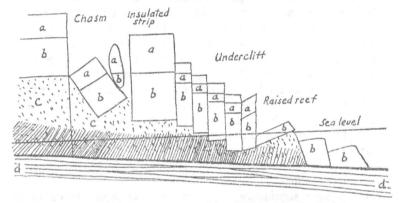

Fig. 8.5 Diagrammatic section of Dowlands and Bindon landslip
(*a*) Chalk. (*b*) Cherty Sandstone. (*c*) Foxmould: lower beds shaded where reduced to quicksand. (*d*) Lias (Gault not distinguished).
After wood-cut in Conybeare's Memoir (1840)

recognised that disturbance of the ground may lead an observer to 'see stars'. A preacher, who evidently regarded an earthquake as a direct manifestation of the power of the Almighty, denounced as infidels all those who believed the landslip to be due to 'secondary' causes; and in a pamphlet published in London by a millenarian, and entitled *A brief Account of the Earthquake, the solemn event which occurred near Axmouth*, it was recognised as the fulfilment of a prophecy in the book of Revelation.

These accounts, together with the exaggerated reports which had appeared in certain newspapers, naturally led to considerable alarm, and intending visitors to Lyme were discouraged by their anxious friends. The greatest danger appears however, to have lain in the impressive nature of the view, for we read that 'many are breathless and bewildered at the sight', while 'one individual from Honiton, was taken home to a sick bed, from which he was with difficulty recovered'. Nevertheless an increasing number of visitors came to the scene, and by charging an entrance fee to their land the farmers of Dowlands and Bindon were soon more than compensated for the loss of their ground. As many as a thousand tickets were sold in one day, steamers brought parties from Weymouth and Torquay, Mrs Critchard returned to her ruined house to supply refreshments, and the idea of the landslip so caught the popular fancy that a dance called the Landslip Quadrille was published with a lithograph of the chasm on the title-page. The excitement culminated in August 1840, when a grand rustic fête was arranged in connection with the reaping of the corn in the isolated fields 'by attendants of Ceres'. The crowd was so great that few people could see the ceremony, and many could not obtain refreshments in spite of the generous supplies of provisions. The 'wagon-loads of delicious food' spread out on the grass, the hams, home-made bread, and water 'boiled in furnaces', are still quoted by Mrs Gapper, who has herself lived for the last fifty years in the cottage rebuilt near the site of Critchard's, and whose mother was among the farmers' wives who reaped the corn in that harvest of 1840.

Whitlands Cliffs

An extensive subsidence, reported to have taken place in the Whitlands Cliffs in 1765, may (Roberts, 1840) have been a result of the 'memorable wet season' of 1764. The excessive rainfall of 1839, which at Christmas had caused the great slip at Dowlands and Bindon, continued until 10 February 1840, and on the morning of the 3rd of that month a further subsidence took place at Whitlands. The main movement was in the undercliff, the lower part of which sank towards the sea, thus exposing a new face of Chalk, a quarter of a mile long and upwards of 60 ft high, within the mass of beds which had fallen on previous occasions. The lower undercliff was rifted and thrust forward, and cottages occupied by coastguards were upheaved and twisted in the same way as those at Dowlands (see Plate 7). The ground on which they stood was tilted, and flooded by streams which had been diverted by the disturbance. Humble Point was forced out to sea, so that it projected sufficiently far to become visible, beyond Seven Rock Point, from the Cobb at Lyme.

A double reef was at the same time elevated on the beach. Soon after Christmas, Mary Anning, the 'fossilist' of Lyme Regis, had pointed out an incipient ridge, 1 ft high, on the shore; this continued to develop, but did not become prominent until the major movements occurred on 3 February. The reefs extended for more than half a mile, the inner one being formed, as at Dowlands, of Upper Greensand (cherty sandstone, Foxmould and Cowstones), 100 ft beyond the old line of sea cliffs. The outer ridge, which was flooded at high tide, was composed of shingle, Cowstones and drift-capped Chalk which had fallen in previous slips and was now again upheaved. Between the two reefs lay a long narrow pool of salt water.

The formation of these reefs was watched by observers, one of whom actually walked along the inner ridge while it was being elevated beneath his feet. He saw the semi-fluid Foxmould being squirted up through rifts among the harder rocks, and thus confirmed Conybeare's hypothesis of the relative roles of the various beds. The phenomena at Whitlands could all be explained (Conybeare *et al.*, 1840) in the same way as those at Dowlands, although the Whitlands reef was formed simply by the pressure of the subsidence in the undercliff, and there was no sinking of the main inland cliff leading to the formation of a chasm. The reefs were again short-lived. In connection with their development at Whitlands, Dowlands and Hooken, it may be noted that the elevation of reefs of Gault, Chalk and sand, 20 ft high and enclosing lagoons, some 70 yds to seaward of the previous edge of the beach, was reported in the subsidence of the Warren at Folkestone, during the very wet spring of 1937.[1] The conditions controlling landslipping in the Warren (Osman, 1917) are, however, very different from those in south-east Devon.

Haven Cliff

The landslipping of Haven Cliff, above the mouth of the river Axe, and of the cliff which continues eastward towards Culverhole Point, is of comparatively recent date. Roberts, in 1840, mentioned that 'some extensive cracks' were to be seen above the harbour, but at that time there had evidently been no serious amount of falling. In this section, Gault, Upper Greensand and Chalk rest directly on about 100 ft of Keuper and Rhaetic beds. The cliffs are almost sheer, but about half-way up the face, at the top of the Trias and Rhaetic, a narrow terrace, covered with Cretaceous debris, forms an incipient undercliff, which is overgrown in Haven Cliff itself. To the east, great masses of Upper Greensand and Chalk have fallen on to the

[1] See *Folkestone Express*, 20 Mar 1937; *Folkestone Herald*, 20 Mar 1937.

shore and become banked up against the cliff. The actual line of the unconformity is hidden by talus, but, presumably, the slipping takes place at the plane of junction between the sands of the Upper Greensand and the argillaceous Gault, Rhaetic and Keuper beds which underlie them.

The cliff top, just west of Culverhole Point, shows considerable fissuring, and falls occur at the present day, but it is not possible to foretell how extensive the landslipping will ultimately become. Haven Cliff and its easterly extension, however, probably exhibit the incipient stages of such a process as has, in the course of centuries, developed the undercliffs between Culverhole Point and Lyme Regis.

IV. THE FUTURE OF THE LANDSLIPS

During the last hundred years there have been no further important landslips in south-east Devon, where it appears that general equilibrium has been attained in the cliffs and undercliffs, except possibly in the Haven Cliff region. On the other hand minor changes continually occur, and above the chasm at Dowlands fissures are gradually opening near the edge of the cliff, indicating that in due course slices of the present cliff face will undoubtedly subside.

Every winter, also, a certain amount of settling takes place in the undercliffs, and in recent years slipping movements have occurred, for instance, among the fallen Chalk near the shore east of Humble Point, and in the low seaward masses of Cretaceous debris at Bindon. These movements are, however, of a comparatively superficial character, and follow the effects of weathering on fallen material which is intrinsically unstable.

Since no part of the Axmouth landslips can be considered secure ground, schemes which have been promoted for their development as a site for holiday resorts or camps seem particularly unsuitable. It is greatly to be desired that the National Trust may be able to acquire any parts of the cliffs and undercliffs which would otherwise be threatened with exploitation.

V. FACTORS CONTROLLING LANDSLIPPING

The distribution of the landslips on the south-east Devon coast depends primarily on the relations of the dips of the beds, and of the unconformity, to sea-level. Where the plane of junction between the Foxmould sands and the underlying clays occurs in the cliff section and slopes down in a seaward direction, erosion, by cutting the cliff face and hence removing the outward support of the beds, will tend to bring about slipping of the upper layers over the lower. This condition is fulfilled in the Hooken Cliff area west of Beer Head, and along the coast from Axmouth to Lyme Regis. At Beer Head itself, and in Whitecliff, between Beer and Seaton, the Triassic–Cretaceous plane of unconformity is almost entirely below sea-level. Here, therefore, marine erosion acts on cliffs formed only of Cretaceous material, and the Chalk falls directly to the shore, instead of being carried forward on a sliding base.

As an undercliff develops on the upper surface of the clays, so the factors controlling the slipping become more complicated. Water is impounded, beds are held up by talus, and there are great accumulations of unstable material. The forms of the undercliffs and of the inland cliffs which back them are, however, principally dependent on the low degree of the dips, and on the coherence of the upper beds which are let down by the foundering layers of sand. In south-east Devon the Chalk and cherty sandstones are massive strata, which may break but do not readily disintegrate. When subsidence occurs, they are therefore capable of sliding bodily, with comparatively little damage, since the dip is sufficiently gentle for the movements not to be shatteringly violent. In this way solid portions of the cliff face have come to be moved forwards into the undercliff, and the unique character of the landslips of this area is due to the existence of the isolated masses and pinnacles, and above all to the great block of land beyond the chasm at Dowlands.

VI. SUMMARY AND ACKNOWLEDGMENTS

The description given of the landslips on the coast of south-east Devon, from Branscombe to Lyme Regis, shows that they occur in connection with the unconformity between the Cretaceous and the underlying Keuper, Rhaetic and Lias beds. West of Beer Head is the Hooken Cliff landslip; the rest constitute a series, the Axmouth Landslips, running from the mouth of the river Axe to the county boundary on the east, and including the cliffs of Bindon, Dowlands, Rousdon, Charton, Whitlands, Pinhay and Ware. The history of the principal slips is recounted from traditions and early records; a particular description is given of the formation of the great chasm and reef at Bindon and Dowlands at Christmas 1839, and the future of the cliffs is briefly discussed.

It is concluded that the primary factors controlling the land-slipping are (1) the relation of the sands and clays to one another and to sea-level, in the cliff section; (2) the coherence of the overlying beds; and (3) the low degree of the dip.

I am indebted to Professor O. T. Jones, F.R.S., for facilities for working out this study at the Sedgwick Museum, Cambridge; and to Major O. Allhusen for giving me access to the cliffs and under-cliffs of Pinhay, including the Chapel Rock. I wish also to express my gratitude to Dr W. D. Lang, F.R.S., and Dr H. Dighton Thomas, for discussion of the landslips and for much helpful information, and to Mr A. L. Leach for his criticism and for redrawing Fig. 8.2.

REFERENCES

BUCKLAND, W. (1840) 'On the landslip near Axmouth', *Proc. Ashmolean Soc.*, I, no. 17, 3–8.

CONYBEARE, W. D. (1840) 'Extraordinary land-slip and great convulsion of the coast of Culverhole Point, near Axmouth', *Edinburgh New Philosophical J.*, XXIX, 160–4. (Reprinted from a letter to Woolmer's *Exeter and Plymouth Gazette*, 4 Jan 1840.)

—— *et al.* (1840) *Ten Plates comprising a plan, sections, and views, representing the changes produced on the coast of East Devon, between Axmouth and Lyme Regis, by the subsidence of the land and elevation of the bottom of the sea, on the 26th December, 1839, and 3rd of February, 1840, from drawings by W. Dawson, Esq. civil engineer and surveyor, Exeter, the Rev. W. D. Conybeare and Mrs. Buckland. With a geological memoir and sections, descriptive of these and similar phænomena, by the Rev. W. D. Conybeare. The whole revised by Professor Buckland.*

DE LA BECHE, H. T. (1830) *Sections and Views, Illustrative of Geological Phænomena.*

—— (1831) *A Geological Manual*, 1st ed.

LANG, W. D., and THOMAS, H. (1936) 'Whitsun field meeting, 1936. The Lyme Regis district: Report by the directors', *Proc. Geol. Assoc.*, XLVII 301–15, Plates 32–4.

OSMAN, C. W. (1917) 'The landslips of Folkestone Warren and thickness of the Lower Chalk and Gault near Dover', *Proc. Geol. Assoc.*, XXVIII 59–84, Plates 5–8.

ROBERTS, G. (1840) *An Account of and Guide to the Mighty Land-slip of Dowlands and Bindon, in the parish of Axmouth, near Lyme Regis, December 25, 1839: With the incidents of its progress, the locality, historical particulars, its causes popularly treated, and the claims of its being the effect of an Earthquake considered with mention of a more recent movement at Whitlands February 3, 1840*, 1st–4th eds, and 5th ed. with illustrations (Lyme Regis).

ROWE, A. W. (1903) 'The zones of the White Chalk of the English coast: III. Devon. The Cliff-sections by C. Davies Sherborn', *Proc. Geol. Assoc.*, XVIII 1–51, Plates I–XIII.

WANKLYN, C. (1927) *Lyme Regis, a Retrospect*, 2nd ed. (A map used in connection with the lawsuit may be seen in the museum at Lyme Regis.)

WOODWARD, H. B., and USSHER, W. A. E. (1911) 'The geology of the country near Sidmouth and Lyme Regis: with contributions by A. J. Jukes-Brown', *Mem. Geol. Surv. England and Wales*, Explanation of Sheets 326 and 340, 2nd ed.

APPENDIX

The Plane of Landslipping on the Coast of South-east Devon

The preservation from possible exploitation of the landslips between Axmouth and Lyme Regis was happily secured in 1955 when they were declared a National Nature Reserve by the Nature Conservancy. I am grateful to the Conservancy for a permit to continue field study in the Reserve, and to Dr W. A. Macfadyen for much discussion of the geological problems involved in the slipping. I am deeply indebted to the Warden, Mr L. A. Pritchard, for his help year after year; under his guidance I have now explored areas which had previously been inaccessible to me, and as a result I have come to see structure in what I formerly regarded as merely fallen and jumbled material in the undercliff.

Although I have not been able to undertake the work of mapping them, I now recognise that the system of ridges and gulleys running parallel to the sea at Pinhay extends as a complex and irregular series westwards as far as Culverhole Point. From the summit of the isolated block at Dowlands and Bindon there are visible three or four such ridges on the seaward side, the smaller being so narrow as to resemble broken masonry.

In my paper on these landslips, published in 1940, and for many years afterwards, I accepted the explanation of the slipping given by Conybeare, and I disputed (1962) the arguments of W. H. Ward (1945) that the slips were due to rotational shear slip, but I have now come after all to believe that rotational shear slipping is probably the major mechanism involved. This would account for the series of ridges running parallel to the inland cliffs from which they could have slid in the distant past, although geological mapping is needed to find out whether each is in fact a block dipping inland. In the section given in Conybeare's memoir, the blocks in Dowlands chasm are shown thus dipping towards the parent cliff. The isolated block, on the other hand, is recorded as having moved seawards and subsided to some extent when the chasm foundered and isolated it in the great slip of 1839, and its present-day surface, which still bears traces of the original vegetation, slopes slightly seawards, so that this mass was not apparently involved in any rotational movement.

Conybeare considered the plane of slipping to be at the junction of the Foxmould and Lias, but on the rotational shear-slip hypothesis it must be postulated as being at some deeper level in the Lias or even in the Rhaetic. Ward (1945) pointed out that this would account for the upheaval of the Greensand reefs at Dowlands and Whitlands in 1839 and 1840. It would also explain an upheaval of Lias clay which occurred in 1961, forming a scarp about 10 ft high on the foreshore at Humble Point. At Pinhay the plane of slipping appears to be more superficial, occurring above the seaward cliffs of Blue Lias; here, in a slip in 1966, a large mud-flow developed in the undercliff above the site of the steps which formerly led down to the beach in Pinhay Bay.

The structural relations of the slipped and solid masses are mostly hidden in the dense vegetation of the undercliffs or else buried far below the surface of the ground. All theories about the planes and mechanism of the landslips are likely to remain speculative until the area can be investigated by borings.

REFERENCES

ARBER, M. A. (1962) in 'Coastal cliffs: report of a symposium', *Geogr. J.*, CXXVIII 303–20.
WARD, W. H. (1945) 'The stability of natural slopes', *Geogr. J.*, CV 170–97.

9 Coral Reefs and Islands and Catastrophic Storms

D. R. STODDART

I. INTRODUCTION

CORAL reefs and islands are distinctive coastal features of tropical seas which have attracted the attention of geomorphologists and biologists since the time of Darwin. The world distribution of these features is reasonably well known (Joubin, 1912; Wells, 1957; McGill, 1958), and the great increase in reef studies since 1945 has enabled us to identify major regional differences in reef character-istics in different parts of the world. Some of these characteristics are probably fossil, and related to either high or low stands of the sea in Pleistocene and Recent times. Others have been interpreted in terms of the normal processes of wind, waves and biological activity, often without specifying the time periods within which these normal processes have been thought to operate.

In recent years considerable attention has been paid to the effects on reef morphology of fairly infrequent events of great magnitude, in the belief that features which cannot be explained by everyday processes may have been caused by major storms occurring at intervals of one to one hundred years, rather than by processes of the geological past which can no longer be observed. The study of such storm effects can not only throw light on the geomorphology of individual reefs, but it may also help us to understand regional differences in reefs on a world scale.

Fig. 9.1 shows the distribution of coral reefs of the world, together with the areas on each side of the Equator where hurricanes originate, and the main tracks of these storms (Bergeron, 1954; Dunn and Miller, 1960). The shaded areas are those where hurricanes fre-quently occur in the coral-reef seas, and where reefs can be expected to show their effects. In the Indian Ocean these areas include the Laccadive Islands, the islands between Madagascar and the Mascarenes, the Chagos Archipelago and Cocos-Keeling. In the North Pacific the reefs of the South China and Philippine Seas, including the Marianas, and the great belt of atolls from the

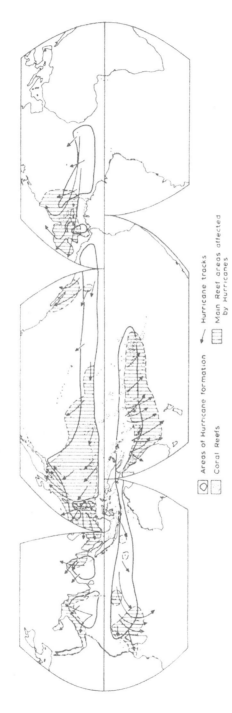

Fig. 9.1 World distribution of coral reefs (Wells, 1957) and hurricane areas (Bergeron, 1954; Dunn and Miller, 1960)

◇ Areas of Hurricane formation   ← Hurricane tracks

▦ Coral Reefs

▦ Main Reef areas affected by Hurricanes

Carolines to the Marshalls are exposed to typhoons. In the South Pacific the hurricane belt includes a large part of the Great Barrier Reef system, and the reefs from New Caledonia and the New Hebrides to Samoa, the Cook Islands and the great atolls of the Tuamotu Archipelago east of Tahiti. All the reefs of the Gulf of Mexico and the nothern Caribbean basin, together with the Bahamas, are subject to hurricane damage.

This list probably includes most of the world's atolls. Many reef areas, however, are not so affected: the reefs of the East African coast and Red Sea, the whole of the Maldive Islands, north-west Australia, and the rich reefs of the Malaysian and Melanesian regions, from Sumatra to the Solomon Islands, and, in the Pacific, the atolls of the Gilberts, the Phoenix group, and many of the Line Islands, together with the easternmost Tuamotus and the Hawaiian Islands. The boundaries of the hurricane-affected areas are of course generalised, and the hurricanes sometimes occur outside them. But within the hurricane zones proper such storms are by no means infrequent: in the western Atlantic alone the number of hurricanes has averaged 4·5 a year since 1900 (Glynn *et al.*, 1965, p. 344), while in one small area, the British Honduras coast, there have been fourteen such storms since 1900, four of them of great severity (Stoddart, 1963, p. 130). It is thus surprising that a recent review of coastal climates does not consider the distribution and frequency of hurricanes (Bailey, 1959; Putnam *et al.*, 1960).

There have been two main opportunities in recent years in the coral seas to make comparative studies of the geomorphic effects of hurricanes by detailed survey before and after their occurrence: one at Jaluit atoll, Marshall Islands, following Typhoon 'Ophelia' in 1958 (Fig. 9.11), the other on the barrier reef and atolls of British Honduras, Caribbean Sea, following Hurricane 'Hattie' in 1961 (Figs. 9.3 and 9.4). Both studies have been followed by later expeditions to study long-term adjustments. Storm effects have also been observed at Low Isles, Great Barrier Reef; at Ulithi atoll, Caroline Islands; at the high islands of Guam in the Pacific and Mauritius in the Indian Ocean; and on the coasts of Louisiana, Florida and Puerto Rico.[1] This paper aims (1) to describe the main

[1] Hurricanes in the United States have mainly affected barrier-beach and salt-marsh coasts, and are not considered in detail here. On the Atlantic-coast hurricanes, see Brown (1939), Howard (1939) and Nichols and Marston (1939); on the Gulf Coast and Florida, see Chamberlain (1959), Engle (1948), Ball *et al.* (1963), Hayes (1966), Morgan (1959), Morgan *et al.* (1958), Oppenheimer (1963), Tanner (1961), Thomas *et al.* (1961), Warnke (1967) and Warnke *et al.* (1966). This

effects of hurricanes in reef areas, with particular reference to the studies in British Honduras and at Jaluit, and (2) to evaluate the long-term significance of such storms in the development of coral reefs and islands.

II. GEOMORPHOLOGY OF REEFS AND ISLANDS

Much early geomorphic work on reefs, following Darwin, concentrated on what may be termed first-order problems: how atolls and barrier reefs came into being, and how their general structure may be explained. These problems lent themselves to deductive geological reasoning, and indeed little can be learned in the field about them without deep drilling and geophysical studies. Hence for many years the 'home study of coral reefs' (Davis, 1914) had preference over field investigation. The study of second-order problems – the detailed morphology of the coral reefs, reef flats and islands, and the processes at work – really began with Finckh's (1904) careful mapping at Funafuti atoll in 1898, but did not become established until the Great Barrier Reef Expedition of 1928–9. Whereas earlier students were most concerned with the geological relationships of reefs on a regional scale, the Great Barrier Reef and most later expeditions have concentrated on reef features measuring from a few centimetres to a few thousand metres in greatest dimension.

Morphology and zonation of reefs
Windward reefs in the Indo-Pacific consist (Fig. 9.2) of a seaward slope, coated with corals on its upper part; a reef-edge zone of encrusting pink algae or boulders; a pavement-like reef flat thinly veneered with sediment, often lacking growing corals; and a lagoon edge and slope coated with coral. The lagoons of both atolls and barrier reefs contain isolated mound-like patch reefs which generally lack a well-developed reef flat. Leeward reefs are lower, and the leeward reef flat may be permanently submerged.

Little is known of the seaward slope on windward reefs, because continuous wave action inhibits direct observation, but from the air the upper slope is seen to be furrowed by grooves and spurs, normal to the reef front, descending to depths of 10 fathoms or more. The occurrence and spacing of these seem to be directly

paper does not deal at all with effects of storms on human geography: in this connection see Bates *et al.* (1963), Lessa (1964), U.S. Army Engineer District (1962) and Yamashita (1965).

related to the strength of the surf; they are not usually found in sheltered areas, or on leeward reefs. The seaward reef edge is marked in many areas by an algal ridge, built by pink, limestone-secreting algae, and standing up to 3 ft above the reef flat. This ridge is well developed in the Marshall Islands, the Tuamotus and the Maldives, but not in the south-west Indian Ocean, the East Indies or the Solomon Islands. Its presence is partly a function of wave strength, but also of world-wide differences in the distribution of the algae. Certainly in many exposed atolls pink algae grow where no corals could survive, because of the intensity of surf, and

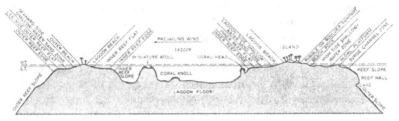

Fig. 9.2   Zonation of an Indo-Pacific atoll rim

it has even been argued that atolls could not exist as surface features were it not for the wave-resistant character of calcareous algae in the surf zone.

Windward reef flats are rock platforms which dry at low water, and which are often so smooth that vehicles can be driven along them. In some areas they are littered near their seaward edge with carpets of rubble and gravel, but otherwise sediment is thin and discontinuous. Corals grow only in deeper pools, and then only hardy species, although the amount of growth increases as the reef flat deepens towards the lagoon. Leeward reef flats, being deeper, generally have more growing coral and a cover of sand. The lagoon-edge reefs are not so well defined as seaward reefs, and lack groove-spur formations, but in the calmer water coral growth is often much more luxuriant than on the seaward side. Coral growth on patch reefs is comparable to that on lagoonward reefs.

Character of Growing Reefs

Coral reefs are constructional features built mainly by the stony corals, a group of sedentary animals which secrete an exoskeleton of calcium carbonate. These skeletons differ widely between species in size, shape and strength, and the animals themselves can be divided into the frame-building and the non-frame-building corals. Frame-builders are either massive, relatively slow-growing, hemispherical forms such as *Montastrea, Favia* and *Goniastrea*, or rapid-growing branching forms such as *Acropora*, and these construct the basic honeycombed framework of the reef itself. The framework shelters many smaller corals, including unattached forms such as *Herpolitha* and *Fungia*, and by growing up into the waves it forms a sheltered zone where delicate branching corals can survive (*Acropora cervicornis* in the Caribbean, *Montipora* in the Pacific). The reef structure serves as an environment for many other reef animals which contribute their skeletons to the sediment accumulating in and around the reef. Carbonate-secreting plants such as *Halimeda* may also contribute skeletal material to the sediment.

The nature of these sediments is closely related to that of the contributing organisms, both in size and in subsequent fragmentation behaviour. Thus the branching tree-like skeleton of some *Acropora* colonies breaks down first into sticks about 6 in. long, and then into sand-size particles. Other corals form plates, slabs or cobbles, but most form first a cobble-size sediment, and then a sand-size. Finer sediments, such as lime muds, may in some cases form by the mechanical breakdown of organic material, but in other cases are formed chemically.

These patterns, which are still very little known, are important in themselves, and also for the effect they have on sedimentary processes and forms on reefs, particularly in the formation of reef islands. There are some indications that the dominant growth forms of corals may vary from region to region in the coral seas. Certainly the Caribbean coral fauna, with its reduced number of genera and species, is much poorer than that of the Indo-Pacific, and many important reef-builders are missing. Within the Indo-Pacific, moreover, the Maldive reefs are remarkable for their large proportion of fragile branching forms and the lack of massive corals. These reefs lie outside the hurricane belt. It is possible that there is a systematic relationship between regional energy environment and dominant coral growth form, similar to that so clearly shown by the zonation of corals on individual reefs.

Sedimentary Reef Islands

Islands in reef areas are of two main types: those formed by the accumulation of reef debris on reef flats, and those formed by mangrove growth. We do not consider here high volcanic islands or islands formed by the relative elevation of reefs above the sea. Islands formed of reef debris can conveniently be divided into two types, motus and sand cays.

Motus (Newell, 1961, pp. 102–3) are characteristic of Indo-Pacific atoll rims. They are long, narrow islands which in some cases may almost entirely encircle the atoll lagoon. The seaward beach may be formed of cobbles and gravel, the lagoon beach is often sandy, and there may be a swampy depression between the two. Occasionally dunes have formed on the crest of the windward beach ridge. The word 'motu' is of Polynesian origin, and islands of this kind have been described from the Tuamotu Archipelago, the Marshall Islands and the Maldives. On some atolls the motus are smaller, but the main characteristics remain: a seaward beach formed by coarse material accumulated close to the seaward reef edge by wave action, and finer sediments accumulating to leeward.

Sand cays are smaller, are formed of gravel and sand rather than boulders and cobbles, and are formed at reef gaps and on reef patches by wave refraction. Typical sand cays have been described from the Great Barrier Reef and from Jamaica by Steers (1937, 1940) and from the British Honduras reefs by Stoddart (1962). Sand cays are also found on some Pacific atolls at lagoon entrances and on lagoon patch reefs, and the distinction between sand cays and motus is by no means sharp.

Once debris has accumulated on reef flats it is stabilised in two ways: by chemical changes leading to the lithification of the sediments, and by vegetation growth. Lithification takes many forms; beachrock is the best known, forming narrow ledges of cemented beach sands outcropping intertidally at the foot of beaches and protecting them against wave attack. Aeolianite (dune rock) and cay sandstone may also form inland beneath the surface of the cay. These processes, by no means fully understood, can form extremely tough and resistant rocks, especially after exposure to the atmosphere, when the friable rock formed by primary bonding is transformed by the filling of voids and by case-hardening.

Vegetation growth is perhaps more important physiographically, since it directly protects the island surface from wind and wave attack, and through the binding effects of roots stabilises loose sediments. Because of their oceanic location, most islands are dependent for their vegetation cover on species which are readily transported by sea, by wind or by birds, and the floras of islands

ACG F

have a considerable number of common elements throughout the Indo-Pacific and to some extent in the Caribbean as well. The strand vegetation consists of pantropical vines, prostrate herbs, sedges and grasses (*Ipomoea, Canavalia, Sesuvium, Cyperus, Sporobolus, Paspalum*) on the upper beach, with a beach-crest shrub community dominated in the Indo-Pacific by *Tournefortia argentea* and *Scaevola*, or by *Pandanus*, and in the Caribbean by *Tournefortia gnaphalodes* and *Suriana maritima*, or by *Thrinax*. Mature woodland is now rare on atolls, having been cleared for coconuts mostly during the last century, and only remnants now remain of mature *Pisonia, Barringtonia, Cordia* and other woodland. It has been replaced by a community of coconuts, shrubs and low trees, with *Guettarda, Morinda* and *Thespesia*, and a rich undergrowth of herbs and grasses, many of them cosmopolitan weeds. Such coconut thicket is perhaps the dominant vegetation on many reef islands today. Elsewhere the natural vegetation has been replaced by *Casuarina* woodland, which forms pure stands with little undergrowth. Man has interfered with island vegetation, by clearing, cropping and introducing cultivated, decorative and weed species, to such an extent that it has been said of the Marshall Islands that the vegetation is no longer a function of species and habitat, but rather of people and plants. These vegetation changes add a further dimension to the problem of hurricane affects, for we must consider not only the relationship between storm effects and native vegetation, but also the effects on the various disclimax communities maintained by man.

Mangrove Islands

Mangroves play an important role on the reefs of the Indian Ocean, the western Pacific and the Caribbean, and also on continental coasts, but because of their unpleasant environment they have been little studied (see Bowman (1918), Davis (1940), Steers *et al.* (1940), Carter (1959) and Thom (1967)). In the Caribbean the pioneer mangrove is the stilt-rooted *Rhizophora*, whose viviparous seedlings float into shallow water and establish themselves on shallow bottoms. The massed roots of *Rhizophora* hinder water circulation and aid sedimentation, and the accumulation of vegetable debris also helps to raise the surface above water-level. Other mangroves then replace the *Rhizophora*, first the black mangrove *Avicennia*, with vertical pointed pneumatophores, then the white mangrove *Laguncularia*, with vertical club-topped pneumatophores, and finally a mature woodland of largely non-mangrove species. Such successions are well documented for Florida and Jamaica, and apply to other parts of the Caribbean. In the Indo-Pacific other genera and species are found. Thus at Aldabra, north of Madagascar, *Avicennia* is found on open lagoon flats, *Rhizophora* on deeper mud in creeks, and *Bruguiera* and *Ceriops* at higher levels at the head of creeks.

Mangrove islands are of two types: patches colonised by mangroves alone, with little or no dry land, and similar mangrove patches with sand ridges banked against their windward sides, bearing a vegetation of littoral woodland or coconuts and strand plants. The classic description of these 'mangrove–sand' cays is that of the Tortugas, off Florida, by Davis (1942). Similar islands are found in British Honduras, where Turneffe atoll, for example, consists of a rim of mangroves with a long sand ridge on its windward side. Mangroves also form an essential part of the 'low wooded islands' of the Queensland coast and of similar islands in other parts of the world.

III. FEATURES OF HURRICANES

Hurricanes are intense cells of low pressure, with a central eye surrounded by a circular wind system, anti-clockwise in the northern hemisphere, clockwise in the southern. The whole system travels at speeds of up to 15 m.p.h., curving away from the Equator as shown in Fig. 9.1. Pressure in the eye may be as low as 27 in., with very steep pressure gradients. Winds in severe hurricanes may reach sustained speeds of 100 m.p.h., with higher gusts; and locations crossed by the eye experience violent winds from different directions separated by a period of calm. The low pressure and the force of the wind help to generate a storm surge which accompanies the storm, and which in shallow coastal-shelf areas may raise sea-level 15–20 ft above normal. On islands rising steeply from the deep ocean floor the surge is much smaller. High wind speeds combined with intense wave action at abnormally high levels are capable of overtopping low reef islands and causing great damage.

Hurricane 'Hattie' can serve as an example of hurricane characteristics. Hattie was first identified as a tropical storm off the east coast of Nicaragua (Fig. 9.3) on the afternoon of 27 October 1961. It moved northwards, and intensified, during 28 and 29 October. Late on the 29th the rate of movement decreased, and the storm began to move westwards towards the Yucatán peninsula. The northern part of the peninsula is low-lying, with a narrow fringing reef; the southern part, adjoining British Honduras, is more diverse, with an extensive offshore barrier reef, and three deep-sea atolls (Turneffe Islands, Lighthouse Reef and Glover's Reef). Gale-force winds extended for 200 miles to the north-east and 140 miles to the south-west of the storm centre as it approached the British Honduras reefs. Between midnight on the 30th and midday on 31 October, Hattie moved across the British Honduras coast and into the Maya Mountains. At Belize City pressure fell rapidly after midnight to a

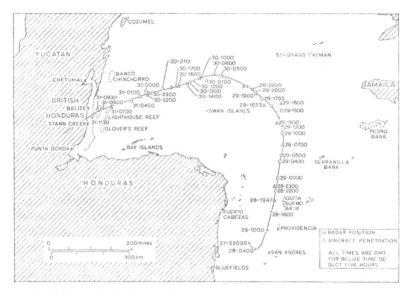

Fig. 9.3   Track of Hurricane 'Hattie' in the Caribbean Sea, 27–31 October 1961 (based on information supplied by Dr Gordon E. Dunn)

minimum at 6 a.m.; the anemometer ceased to function, but reliable estimates put sustained wind speeds at 140–160 m.p.h., gusting to over 200 m.p.h. As the centre of the storm passed, the sea-level rose and flooded the coastal area to a maximum height of 15 ft. The rather limited data suggest that the surge was greatest to the north of the storm track, and higher on the barrier reef and coastal shelf than on the outlying atolls. We have no direct information on wave conditions, since the storm occurred during the night, but such intense wind speeds coupled with the storm surge must have led to very confused and turbulent sea conditions. The storm was accompanied by heavy rainfall, which caused mainland flooding for several days afterwards. Run-off may have subsequently diminished salinity in nearshore areas.

Such conditions, which are comparable to those of other severe hurricanes and typhoons, cause geomorphic and vegetational changes as a result of (*a*) wind, (*b*) waves and (*c*) inundation. Using Hurricane 'Hattie' as an example, we consider now the effects of such storms on (*a*) reefs, (*b*) reef islands, (*c*) land vegetation and (*d*) mangroves. Hurricane effects varied, as the storm passed, with location relative to the centre. North of the storm track, winds were first northerly, then backing to north-east and east, and increasing. In this sector the winds were blowing in the same direction as the storm movement, and both storm surge and waves were at a maximum. South of the storm track, winds were westerly, veering south and south-east as the storm passed; they blew in the opposite direction to the storm movement, and hence off the land. Because of restricted fetch and shallow coastal water, seas were lower, but interference with waves generated in the northern sector led to disturbed conditions. Areas close to the storm track experienced rapid reversal in wind and sea conditions.

IV. EFFECTS ON REEFS

The centre of 'Hattie' passed north of Lighthouse Reef, across the northern part of the Turneffe Islands, and across the barrier reef and coastal lagoon between Rendezvous Cay and Mullins river (Fig. 9.4). The barrier reef itself in this area is backed by numerous patch reefs up to a few hundred yards in length, and before the storm these had maintained vigorous communities dominated by branching *Acropora* (*A. cervicornis, A. palmata*) and some globular corals (*Montastrea, Diploria, Siderastrea*). From the sea and from the air the edge of the barrier and the surface of the patch reefs had the rich orange colour of growing corals. After the storm the barrier reef and patches between English Cay and Rendezvous Cay were swept clean of living corals; the orange colour gave way to greenish-white. On the barrier reef all trace of the groove-spur pattern was removed, suggesting that the spurs were simply coral-growth features, rather than gaps between grooves eroded into solid rock. Only at Cay Glory, 10 miles south of the storm centre, could the first scattered remnants of the lineations be seen, although coated by dead, not living, corals, and spurs with living coral were not seen within 25 miles of the storm track. Similar massive damage

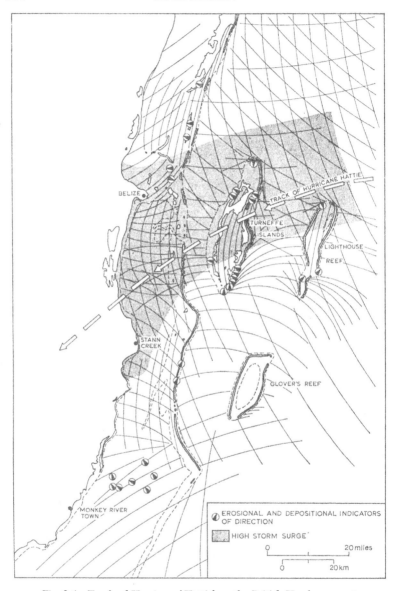

Fig. 9.4   Track of Hurricane 'Hattie' on the British Honduras coast

occurred on the windward side of Turneffe, especially at the northern end, but to a much lesser extent on Lighthouse Reef, which was entirely south of the storm track.

Detailed reef studies had been carried out before the storm on the Rendezvous Cay reef patch, and it was thus possible to trace in detail the storm effects (Fig. 9.5). The only coral to survive around the patch was *Montastrea annularis*, which grows in massive hemispherical colonies; perhaps half was killed. The fragile branching colonies of *Acropora cervicornis* (stagshorn), once one of the most prolific corals, disappeared entirely, and only one-fifth of the more massive branching elkhorn, *Acropora palmata*, survived. A few minor corals were found shortly after 'Hattie', but almost all the prolific branching and unattached forms have been destroyed, and the surface of the reef and its slopes were mantled with rubble.

In Australia the genera and species of the corals are different but the growth forms are analogous. After the 1954 cyclone at Low Isles, Stephenson *et al.* (1958) found that the most resistant species had been the massive ones (*Goniastrea pectinata*, *Playtoyra landlina*, *Porites lutea*), and the least resistant the fragile branching forms (*Montipora divaricata*, *Pocillopora damicornis*). In Puerto Rico the minor Hurricane 'Edith' of 1963, with winds of less than 50 m.p.h., caused extensive coral destruction, especially in the branching corals such as the *Acroporas*. Again the massive forms survived most successfully (Glynn *et al.*, 1965).

Destruction is mostly mechanical; colonies are uprooted, carried above sea-level or more often into deep water, or are fragmented *in situ* by direct wave action. At Rendezvous Cay many colonies were found, especially of finger corals, which appeared unharmed but fragmented at a touch. Other corals are killed as bigger corals roll over them, and smaller globular forms can be rolled long distances. In some cases corals may survive, but mortality is simply delayed. Moorhouse (1936) records how the 1934 cyclone at Low Isles so altered the geomorphology that tidal flow changed and water-level fell, so exposing *Pocillopora* and *Montipora*, and promoting abnormal and ultimately death in *Porites*. It is also possible that corals in certain areas may be killed by the rapid fall in salinity caused by torrential rainfall and river flooding associated with tropical storms. Thus the cyclone of 1918 in southern Queensland, where heavy rain occurred with low tides, destroyed one of the most flourishing reefs on the coast, the damage extending for 20 miles (Hedley, 1925). Goodbody (1961) found massive mortality

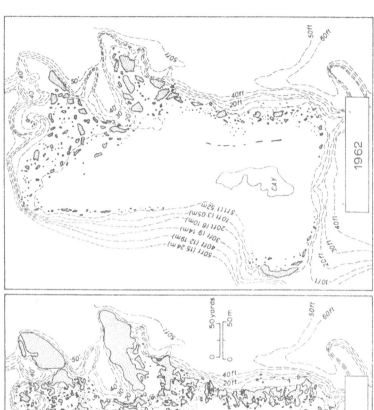

Fig. 9.5 Distribution of corals on Rendezvous Reef in 1960 and 1962

of marine fauna after heavy rain in Kingston Harbour, Jamaica, and following Hurricane 'Flora' in 1963. Goreau (1964) found wholesale bleaching of corals in shallow-water reef communities through expulsion of zooxanthellae or symbiotic algae. On this occasion salinity at Port Royal fell to 3 parts per mille two days after the storm, and remained below 30 per mille for five weeks afterwards. The corals most affected were *Montastrea* and *Millepora*, together with zoanthids and actiniarians. The expulsion, which is similar to that resulting from other kinds of physiological stress, appears to be caused directly by lowered salinity. The corals were not killed, but remained bleached for several months. Since zooxanthellae play a major role in calcification in the corals, the rate of skeleton formation must be generally slowed by this process.

Local differences in exposure can, however, lead to the survival of areas of fragile corals even in extreme storms. At Jaluit atoll, where the quantity of rubble suggested massive destruction of corals on the seaward reefs, delicate corals survived in the lagoon (Blumenstock, 1958*b*, p. 1269; Blumenstock, 1961, pp. 75–8).

In addition to the effects on living corals in British Honduras, some superficial scour holes and channels were cut in reef-flat deposits, and in one area of silty mangrove mud rotational slumping of the bottom deposits occurred. But all of these changes were minor; the major underwater result of severe hurricanes is to remove growing corals over several miles of reef, to destroy selectively the fast-growing branching species over a much wider area, to interrupt active reef growth for an undetermined period of time, and to provide large amounts of coarse sediment, both for accumulation on the reef flats and islands, and to add to sedimentary aprons at the foot of reef slopes. These changes obviously affect the functioning of the whole reef ecosystem; many changes take place in other groups of plants and animals (e.g. Thomas *et al.*, 1961, Tabb and Jones, 1962), but are of less direct geological importance.

V. EFFECTS ON SEDIMENTARY ISLANDS

Zonation of Change

The mapping of reef islands damaged by Hurricane 'Hattie' in British Honduras suggests three main principles of morphologic change during such storms: first, damage is distinctively zoned away from the storm centre; second, damage is greater on small than on large, and on narrow than on wide islands, at any given distance from the centre; and third, damage is more intense on islands stripped of natural vegetation, or where vegetation has been much altered by man.

Zonation of damage is shown in Fig. 9.6, the zone of maximum or catastrophic damage extending for 15–20 miles north and south of the hurricane track. Over the greater part of this area sea-level rose considerably, and winds throughout the zone probably reached sustained speeds of 150 m.p.h. A number of small sand cays dis-

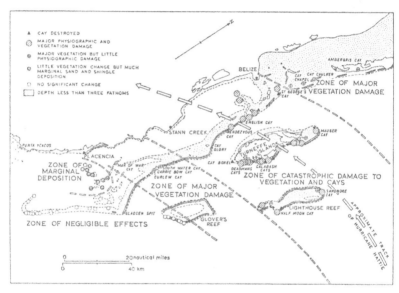

Fig. 9.6   Zonation of damage to sand cays caused by Hurricane 'Hattie'

appeared in this zone; some were completely stripped of larger plants; the geomorphology was in all cases much altered, mainly by marginal erosion and the stripping of surface sand and cutting of channels by overtopping water; and great damage was caused to human habitations, lighthouses and jetties. Goff's Cay, for example, on the barrier reef, was a small sandy island, less than 100 yds long, covered before 'Hattie' with coconuts, sea grape and small black mangrove, and with a ground vegetation of herbs, vines and grasses. During the storm all the trees disappeared, and the island decreased in area by about 60 per cent. The remnant of the original cay remaining had a surface of coconut roots from which all the surface sand had been stripped, and steeply undercut margins. Soon after

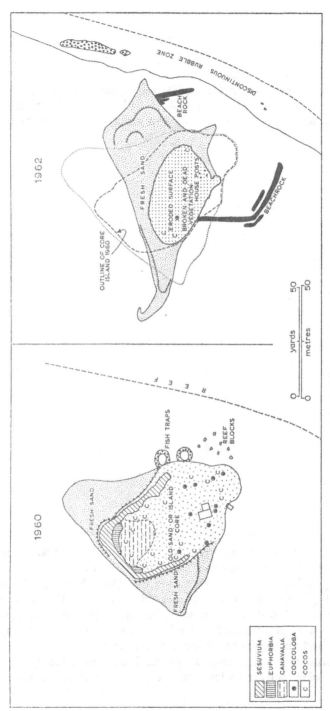

Fig. 9.7 Goffs' Cay in 1960 and 1962

the storm fresh sandpits were built on to this remnant, but five months later the only plant coloniser was *Portulacca oleracea*, growing in scattered patches (Fig. 9.7). Soldier Cay (Fig. 9.8), on Turneffe Islands, was a larger island with a seaward shingle ridge and a dense cover of coconuts. All vegetation was swept away,

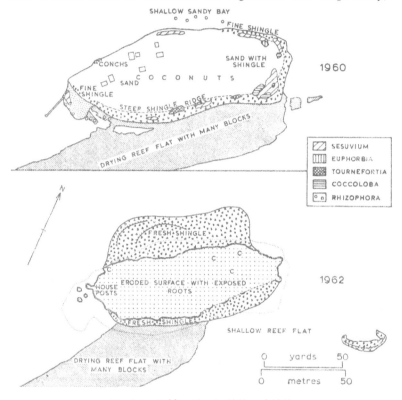

Fig. 9.8 Soldier Cay in 1960 and 1962

except for four coconut trees and some grasses. Sand was scoured from the old cay surface to a depth of 2 ft, leaving a barren surface of exposed roots, and the shores were cut back 5–10 yds. A carpet of shingle was deposited on the lee side of the cay. The old shingle ridge stood 5–6 ft above the sea; after the storm the highest point, on the new gravel spread, was only 3 ft. Seven people sheltering on the island during the hurricane were drowned.

North and south of this central area is a second zone, 15 miles or more wide, subject to less extreme though still violent wind and wave conditions, but largely unaffected by the storm surge. Geomorphic changes in this zone were generally minor, being restricted to shoreline retreat and cliffing and some nearshore sand-stripping and deposition. Vegetation suffered heavy damage, and on many islands the dominant hurricane effect was tree-fall. Thus at Tobacco Cay (Fig. 9.9), on the barrier reef south of the storm track, physical changes were limited to minor retreat along the east and south shores, with stripping of surface sand and exposure of roots near the beach, followed by the deposition of a carpet of fresh sand up to 2 ft thick and 15 yds wide along the south and west shore (dominant winds here were south and southeasterly). By contrast about 70 per cent of the coconuts were felled, especially on the southern half of the island.

A third zone includes the cays between Placencia and Gladden Spit, in the central barrier-reef lagoon, where the main hurricane winds blew from the south with restricted fetch across water 15–25 fathoms deep. On most of these islands the vegetational effects were insignificant, but all had sand or shingle deposited on their south and east shore. At Cary Cay (Fig. 9.10) the fresh shingle ridges extend for 400 yds along the east shore, and consist mainly of *Acropora cervicornis* sticks. This zone lies 30–40 miles south of the hurricane track; it is not duplicated to the north in British Honduras waters, where the barrier-reef lagoon is very shallow and the cays few in number; to this extent it depends on local conditions.

Finally, the cays of Glover's Reef and the southern barrier reef suffered little or no geomorphic change, apart from insignificant shoreline adjustments which may have been partly a seasonal phenomenon. The zone of no geomorphic change extends outwards from a distance of 40 miles from the hurricane track.

Types of Geomorphic Change: Erosional Effects
The main types of destructional changes caused by Hurricane 'Hattie' can be grouped as follows:

1. *Destruction of cays.* A number of cays disappeared completely in Zone I, including Paunch Cay, St George's East Cay, Cay Glory, Cay Bokel, Big Calabash East II Cay, Blackbird Cay, some of the Cockroach Cays and Saddle Cay. All of these except Paunch Cay were vegetated, although most had only coconuts and ground vegetation. Most were low-lying and sandy, and all were small. The largest to disappear was 120 yds long.

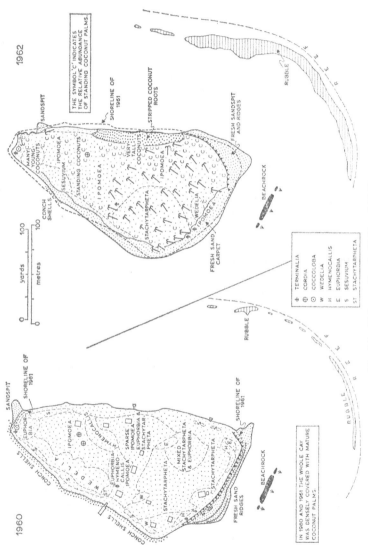

Fig. 9.9 Tobacco Cay in 1960 and 1962

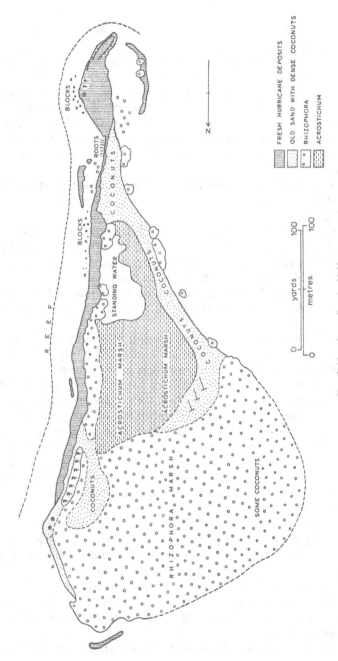

Fig. 9.10 Cary Cay in 1962

2. *Marginal erosion.* Retreat of cay shorelines took place in Zones I, II and III. All material above sea-level and for several inches below it was eroded away in a zone up to 20 yds wide, generally wider near the storm centre and on the side of the cay facing hurricane winds. This left a vertical cliff, usually 1–2 ft high, capped by a mat of coconut roots from which all sand was flushed. Often the presence of coconut boles gives this cliffed shore a scalloped outline, with small promonotories and bays. Occasionally residual parts of the old cay have been isolated by marginal erosion of this type.

3. *Destruction of unconsolidated spits.* Seasonal spits of un-vegetated sand were washed away, but soon re-formed in the same positions.

4. *Channel-cutting.* Cutting of channels through cay surfaces was limited to Zone I, in the area of the high storm surge, and to cays with open vegetation. Scouring of deep channels occurred at St George's Cay, and shallower channels were cut through narrow necks of land at Mauger and Sandbore Cays. Incipient channels, cut back from the lee side of cays by headward erosion, were seen on many islands.

5. *Scour-holes on cay surfaces.* Erosion of scour-holes by over-topping water was widespread in Zones I–III, especially near obstacles to water flow such as buildings or tree trunks.

6. *Stripping of surface sand.* Where islands were overtopped by the storm surge in Zone I or had their margins submerged by heavy wave action in Zones II–III, stripping of loose surface sand was almost universal, although confined to a zone up to 30 yds wide immediately inland from the undercut cliffline. The loss of sand was only a few inches, but sufficient to expose tough coconut root-mats. In Zone I the root-mat could be several inches deep and devoid of soil; elsewhere it was much thinner and overlay tightly packed sand.

7. *Erosion of consolidated deposits.* Beachrock was very successful in resisting storm damage, though in places large blocks were torn loose and thrown on to the beaches. Much poorly cemented in-cipient beachrock disappeared with beach retreat.

8. *Uprooting of trees,* chiefly coconuts, in Zones I–II formed depressions up to 3 yds in diameter and 3 ft deep, out of range of active water erosion. Many such holes reach the water table and form small ponds.

9. *Beach movement.* Where high beaches are developed, erosion

on the beach face together with deposition on the beach result in shoreward migration of the beach. This is well seen at Half Moon Cay, discussed below.

Types of Geomorphic Change: Depositional Effects

Depositional effects following 'Hattie' were largely confined to the islands and their margins. In spite of the great reef destruction little material accumulated on the reef flats above water-level, in the form of boulder ridges or reef blocks. Surprisingly, stranding of reef blocks only occurred in two places, where transport across the flat was inhibited by islands, though it has often been stated that hurricanes are necessary agents for the lodging of such blocks. 'Hattie' was apparently so severe that material carried on to the reef flats was transported across them into deep lagoon water, or on to island shores, or more probably was carried down the steep seaward reef slopes directly into deep water. Rubble carpets, often of imbricate coral fragments, are common in shallow water where growing reefs have been destroyed, but they rarely rise above the surface. Depositional cones of sand are found at the outlets of erosional channels, and shoaling occurred on the leeside of heavily eroded cays. The main depositional effects are, however, subaerial. They are not well developed on the barrier reef in Zone I, where erosion dominated, but are found in the same zone on the east side of Turneffe, where the islands are longer and more nearly resemble Pacific motus; they are not typically developed in Zone III, where erosional effects are minimal. As will be seen, the nature of the vegetation is critical in distinguishing eroding from aggrading islands in storms.

Subaerial deposition takes the following forms:

1. The littering of heavily eroded surfaces with coarse coral rubble, as at Cockroach Cay and Sandbore Cay, in Zones I and II.

2. The accumulation of coral gravel and sand against vegetation barriers, for example at Cockroach II and Half Moon Cays, where debris is piled up to a height of 10 ft above sea-level, and on several of the central barrier-reef cays (Zone IV), such as Owen and Laughing Bird Cays.

3. Deposition of wider, thinner carpets of sand on the old cay surface, especially inland from the marginal erosion zone already described, in Zones II and III. This is well seen at Tobacco, South Water, Trapp's, Scipio and Colson Cays. The carpet may be up to 30 yds wide, wedging out seawards and thickening landwards,

terminating on the landward side in a steep face, often arcuate in plan. The sand buries the old cay surface, with its soils and vegetation, to depths of up to 2 ft. Even where the sand is thickest, the tall vegetation – especially *Thrinax* – may protrude through it and survive.

4. Deposition of shingle ridges around old island shores, following shore retreat. These are typically developed in Zone III, where the ridges may or may not directly adjoin the shore for their whole length, or may enclose small pools. The ridges are narrow and generally less than 3 ft high.

5. Extension of leeward shores of islands by sand deposition: this occurred on a few islands and on a minor scale.

Incidental changes

The widespread changes in beach location during the hurricane exposed many areas of lithified island sediments. In the case of intertidal beachrock the hurricane re-exposed old rock which had been buried by later sediments, revealed stretches of well-cemented beachrock which had previously been noted as putty-like incipient beachrock, and exposed massive beachrock the existence of which had not previously been suspected. Much of the newly exposed rock is loosely bonded and friable, particularly in coarser sediments. Clearly beachrock is forming under many more beaches than is suggested by its normal exposure; all the outcrops exposed by 'Hattie' are intertidal or slightly subtidal, and show characteristic seaward dip.

Other rocks exposed stand well above high water, on the face of retreating beaches, especially on the eastern sand ridge at Turneffe, and on some of the larger mangrove–sand cays of the barrier. These exposures are again lightly cemented at first, and contain coconut roots and human artefacts. They are a water-table phenomenon not directly related to sea-level, and certainly do not indicate relative movement of land and sea. Some of these outcrops are of coarser gravel rock, in which the cement may be phosphatic. More enigmatic are exposures of low promenades at or slightly above high water, lacking the structural features of true beachrock, lightly cemented, but similar to many toughly cemented platforms of reef sediments which have been used elsewhere as arguments for recent changes in sea-level. If it can be shown that these rocks, formed beneath cays by processes not fully understood, and exposed by major storms, can survive normal wave action and become lithified sub-aerially, then we have a simple explanation for features from which important historical deductions have been drawn.

Geomorphic effects of Hurricanes in other regions

The geomorphic effects of Hurricane 'Hattie' are similar to those noted on small reef islands elsewhere, mainly on the Australian barrier reef (Moorhouse, 1936; Gleghorn, 1947). At Jaluit atoll, Marshall Islands, where Typhoon 'Ophelia' crossed the centre of the atoll in January 1958 (Fig. 9.11), the islands are longer than the Caribbean sand cays, and present a greater barrier to cross-reef water movement. At Jabor and Jaluit islands, about 7 miles long, rubble bars were formed on the seaward reef flat, up to 8 ft high, 15–25 yds wide and 15–30 yds offshore. They consisted of coral cobbles and gravel, with frequent boulders 1–5 ft in diameter. The seaward shores retreated up to 10 yds, and in places new beach ridges were built, similar to the offshore gravel tracts in all features except location. Overtopping water stripped fine sediments from the island surfaces, which were covered by blanket deposits of gravel a few inches thick. This was well seen at Jabor and Mejatto islands, which were flooded, but Pinlep and Majurirek, which were not, retained their original soil cover. The fine material stripped from the island surfaces accumulated to form lagoon bars up to 100 yds offshore, and poorly sorted submarine deposits on the lagoon slope and lagoon floor. Erosional features on island surfaces included scour-holes and channels, which on narrower islands cut completely across the surface. The effects at Jaluit differ from those in British Honduras mostly in the importance of the offshore gravel bars, resulting from the barrier to cross-reef water movement, caused by the linear character of the reef islands. Such gravel-bar formation may be normal on Indo-Pacific atolls during storms (see the accounts of Jaluit atoll by Wiens and McKee in Blumenstock (1961) and McKee (1959)).

In Mauritius, where the storm surges during Hurricanes 'Alix' and 'Carol' were low, rising seas resulted in beach erosion and falling seas in deposition. McIntire and Walker (1964) find a direct correlation between beach erosion and reef-flat width, and find that beach height in relation to storm-wave height was an important control of erosion and deposition. Sediment from eroded beaches was either carried on to the berm, where it becomes a source for dune sands, or on to the reef flat, where with the material from the seaward slopes it forms a reservoir of future beach sediments.

D. R. Stoddart

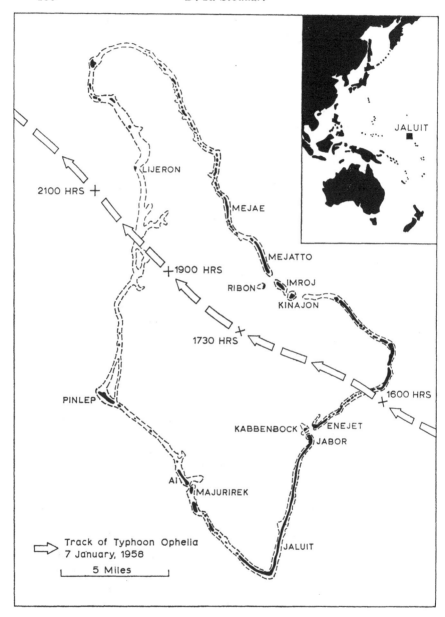

Fig. 9.11   Track of Typhoon 'Ophelia' at Jaluit atoll, 7 January 1958

Storm changes on beaches backed by solid rock, both in Mauritius and on high islands such as Guam (Emergy, 1962, pp. 59–61), are minor compared with coral islands.

In no case of observed storm damage have really large reef blocks been carried on to reef flats. At Jaluit and at British Honduras the largest individual coral colonies rolled on to flats were about 5 ft in diameter, and apart from beachrock no consolidated portions of reef rock were torn from the reef. The largest storm blocks recorded are those at the north-east part of Rangiroa atoll, Tuamotu Archipelago (Stoddart, 1968*a*), where highly cemented blocks of reef rock up to 20 ft in height are jumbled on the reef flat. These are thought to date from a storm in about 1900, when conditions must have greatly exceeded those of any recent hurricane in violence.

VI. EFFECTS ON ISLAND VEGETATION

Damage to vegetation on the British Honduras cays during Hurricane 'Hattie' depended on local storm conditions (largely a function of distance and direction from the storm centre), whether the vegetation was affected by direct submergence and wave action or by wind alone, the type of the vegetation, the species involved, and the nature and behaviour of the substrate. It is convenient to consider the effects of those variables in terms of vegetation types:

1. *Strand vegetation* consists of prostrate vines and creepers, low herbs, and some grasses and sedges (Sauer, 1959). Dominant species include *Ipomoea pes-caprae, Sesuvium portulacastrum, Euphorbia mesembrianthemifolia, Cakile lanceolata* and *Sporobolus virginicus*. Damage, as Sauer (1962) found in Mauritius, largely depended on the degree of hurricane modification of the substrate itself, and to a lesser extent on direct affects on plants. Since many beaches moved landwards, the vegetation on them was obviously destroyed. Strand vegetation on exposed beaches in Zones I–III was therefore damaged, the destruction extending right across the island in the case of small cays in Zone I. Where inundation occurred without beach erosion, grasses survived better than *Ipomoea* and *Sesuvium*, which were easily swept away.

2. *Beach-crest shrub community* consists of shrubs and low trees separating the strand community from the interior woodland or coconuts. In the Caribbean *Tournefortia gnaphalodes* is common on exposed shores, especially on shingle, and *Suriana maritima* on more protected, sandy shores. In Zone I, where the rise in sea-level and violence of wave action reached a maximum, this vegetation

type has disappeared. Relocation of beaches destroyed all mature *Tournefortia* and *Suriana* on Lighthouse Reef, Turneffe Islands and the northern barrier reef. Damage in Zones II and III was much less, but even at South Water Cay, where beaches suffered little change, *Tournefortia* and *Suriana* markedly decreased as a result of wave action. Shoreline stands of the palmetto, *Thrinax parviflora*, proved very resistant to storm damage.

3. *Interior woodland.* Mature woodland has been cleared from most islands for coconuts, and thickets of *Cordia sebestena, Bursera simaruba, Ficus* and other trees now only occur in a few places. In Zone I damage to this woodland was of two types: direct wave damage around the margins, and wind damage in the interior. The first, often associated with relocation of beaches, led to uprooting of trees or stripping of sand from their roots. *Bursera* and *Coccoloba* retained root-holds in these circumstances better than *Cordia*, but often lost leaves and branches. At the inner limit of wave action on larger islands, trees were buried by fresh gravel deposits. The interior part of the woodland which escaped wave action suffered mechanical wind damage, with loss of many branches and defoliation; at Half Moon Cay, for example, the height of the woodland canopy decreased by half. In Zones II and III it is difficult to evaluate hurricane effects on woodland because of its poor development, but judging from individual species damage in Zone III was negligible.

4. *Coconut plantation and thicket.* Coconuts are now the dominant species on sand cays, having been spread by man in the last two hundred years. Coconuts were totally destroyed where cays were washed away during the storm, and wherever shorelines shifted position. Post-hurricane shores in Zones I–III were generally lined by fallen trees. Away from the effects of wave action wind felled approximately 75 per cent of all trees in Zone I and at least 50 per cent over most of Zone II. The direction of tree fall shows a good correlation with that of first hurricane-force winds, rather than, as Wiens suggested at Jaluit (Bluemnstock, 1961, p. 21), with most intense hurricane winds. In Zone I tree fall occurred, either by uprooting or snapping above ground-level; many trees stood but lost their crowns. In Zones II–III uprooting was most evident along shorelines. Heavy damage to coconuts extended further south than north of the storm track (e.g. Tobacco Cay, Fig. 9.9).

Underneath the coconuts there is a ground cover of grasses, herbs and creepers, including common weeds (*Stachytarpheta jamaicensis, Wedelia trilobata, Euphorbia* sp.), with low bushes (*Ernodea littoralis, Erithalis fruticosa, Rivina humilis*) and the spider-lily, *Hymenocallis littoralis*. Near cay shores, and on surfaces inundated because of either erosion or deposition of sediments, this ground vegetation was destroyed. On larger islands in Zone I this peripheral destruction reached a width of 60 yds, but was generally less than 20 yds in Zones II and III. Outside the storm-surge area the ground vegetation suffered no major changes.

In some degenerate coconut plantations, shrubs and other trees combined with coconuts to form a coconut thicket. While the coconuts themselves suffered heavy damage, the rest of the thicket behaved rather like interior woodland (type 3).

Vegetation damage during 'Hattie' was thus caused by inundation, direct wave action, the erosion of the substrate or burial by fresh deposits, and the effect of spray-laden winds. As with geomorphic changes, the effects of these factors are often controlled by local conditions of size, aspect and location. The probability of major vegetation destruction is greater the less dense the vegetation, the lower and smaller the cay, the higher the storm surge, and the greater the wave and wind action. The kinds of damage suffered – uprooting, removal of branches, defoliation, decapitation – also varies with species, being noticeably greater in introduced species such as the coconut, and less in native woodland trees such as *Cordia*. Clearly the interlocks between vegetation growth and sedimentation already mentioned mean that the behaviour of different vegetation communities is intimately related to geomorphic changes, and that the two cannot really be discussed independently of one another. The implications of these interlocks are discussed in section VIII.

Effects on vegetation of hurricanes in other regions
The patterns of damage at Jaluit atoll in 1958 (Plates 9–12) were comparable to those in British Honduras, though many of the species present were different. The islands were covered with plantations of coconuts and breadfruit, with a beach scrub of *Scaevola sericea* and *Tournefortia argentea*. Fosberg (in Blumenstock, 1961) found that damage was greatest on the eastern side of Jaluit, and on narrower islands. Where flooding occurred, with erosion or deposition, ground vegetation suffered, but outside this zone there was little damage. Coconuts and breadfruit underwent major

physical damage, and salt contamination of ground water helped to kill the breadfruit. *Pandanus*, forming an understorey on the plantations, was much broken. Fosberg considered that *Pemphis acidula, Cordia subcordata, Calophyllum inophyllum* and *Casuarina equisetifolia* survived best, and *Pandanus*, breadfruit, *Terminalia catappa* and coconuts worst. Fosberg's photographs of damage to coconuts are comparable to the effect on cays in Zone I in British Honduras; and in the case of *Pandanus*, one of the most common trees, he found the 'trees generally very seriously battered, some uprooted, many more broken off between stilt roots and first branches, or with most of the branches broken off . . . either broken on leafless part or torn from trunk' (in Blumenstock, 1961, p. 57).

The observation that *Casuarina*, a widespread Indo-Pacific tree of the cedar family, survived well at Jaluit, contrasts with the intense damage that *Casuarina* plantations suffered around the Mauritius coast in 1960, especially in the case of trees less than ten and more than thirty years old (Sauer, 1962). Otherwise the Mauritius observations on strand and beach-shrub vegetation and coconuts are in agreement with the atoll data. Unpublished observations of hurricane damage at Guam by Fosberg and Sachet and at Ulithi atoll, Caroline Islands, by Blumenstock *et al.* give further confirmation of the general pattern.

VII. EFFECTS ON MANGROVE ISLANDS

Among the most remarkable vegetational effects of Hurricane 'Hattie' was that on mangroves, particularly on the mangrove rim of Turneffe and on the large mangrove islands of the coastal shelf. Before the storm the peripheral rim of stilt-rooted *Rhizophora* had a rounded outline and rich green colour, with the crowns of *Avicennia* and other mangroves appearing inland. In Zone I both nearshore and interior *Rhizophora* was exposed to inundation and severe wave action; elsewhere, interior *Rhizophora* was affected by spray-laden wind alone. Throughout Zone I mangrove was completely defoliated, damage being greatest on windward shores and on small islands. In the interior parts of Turneffe defoliation was less severe, and some trees still had leaves after the storm. It is likely that inundation and direct wave action is a more efficient defoliator than wind alone.

The beginnings of leaf regrowth on the leeward sides of cays (i.e. the west side to the north and the east side to the south of the storm track) was noted in April 1962 at points north of St George's

Cay, and at Southern Long and Cross Cays, respectively 12 miles north and south of the storm track. The zone of massive defoliation was thus about 25 miles wide. A high proportion of mangroves escaped defoliation at distances of 30–40 miles from the storm centre. Individual trees showed extraordinary stability, maintaining a root-hold even when devoid of leaves and with broken trunks The outline of mangrove islands was thus little changed, but from the air the islands looked bare, reddish-brown in colour, and large stretches of mud and water were exposed, which were previously hidden by the canopy.

Damage to the dry-land communities of small sandy areas on mangrove cays – with coconuts, *Borrichia*, *Batis*, grasses and sedges – was similar to that on sand cays.

Defoliation and mechanical damage, resulting in death, have also been studied in the mangrove swamps of the Florida coast (Davis, 1940, pp. 367–8), particularly following Hurricane 'Donna' in 1960 (Craighead and Gilbert, 1962; Craighead, 1964). In the worst-affected areas 50–75 per cent of mature mangroves, some 80 ft tall and 2 ft in diameter, were killed, and smaller trees were defoliated and in some cases uprooted. Defoliation of isolated trees occurred even in the minor Puerto Rico storm of 1963 (Glynn *et al.*, 1965).

In contrast to other kinds of vegetation destruction, the death of mangroves does not release a surface for fresh colonisation, but results in a dense impenetrable tangle of immensely tough trunks and branches that in the absence of fire can survive for years. In Florida in 1960, dead trees from the 1935 hurricane were still standing.

VIII. LONG-TERM SIGNIFICANCE OF STORM EFFECTS

It is clear that hurricanes cause immediate and drastic changes in the geomorphology of coral reefs, and the morphology and vegetation of reef islands, and such changes are amply documented by the Jaluit atoll and British Honduras studies. Both these storms are recent, however, and little time has elapsed for observing long-term readjustment and recovery in reef areas affected by the storms. Hurricane effects were resurveyed three years after 'Ophelia' at Jaluit (Blumenstock *et al.*, 1961) and three years after 'Hattie' on the British Honduras reefs (Stoddart, 1965, 1968*b*), and it is hoped that resurveys can continue at least until the next hurricane strikes.

Much information is also available from other reef areas where storms of known date occurred at some time in the past, and where recovery has been more or less complete.

Recovery of coral reefs

Reef recovery in British Honduras has been slight. In the area of greatest reef damage, the only corals living in any quantity in 1965 were those which survived the storm itself, chiefly *Montastrea annularis*, and in places *Siderastrea, Diploria* and massive *Acropora palmata*. Recolonisation has been limited to a few scattered *A. palmata* less than 30 cm. tall, *Millepora*, and minor non-frame-building corals. Formerly widespread and important species such as *Acropora cervicornis* are still rare or absent in the most severely damaged areas, and wide areas of rubble and dead coral heads are thickly blanketed by *Padina, Halimeda* and other algae. Sponges and gorgonians, while not numerous, have made a more rapid recovery than the corals.

Why reef recovery is delayed is not clear. In Australia, Stephenson *et al.* (1958, p. 304) suggested that the movement of rubble after the storm could not only injure more corals but could also inhibit colonisation and regrowth, and that only the complete flushing away of debris by a major storm would allow redevelopment. 'Hattie', however, simply produced more debris. Competition from vigorously growing algae may also delay growth, as may increased turbidity of reef waters. In these circumstances the estimate made by Stephenson *et al.* (1958, p. 261) of ten to twenty years being required for reef recovery at Low Isles, Great Barrier Reef, is probably minimal. On the other hand the British Honduras reefs, flourishing in 1959–60, had certainly been damaged by the 1931 hurricane, suggesting a recovery period of not more than thirty years.

Some of the dead reef debris has been moved on to reef crests and island shores to form rubble ridges since 1961, but otherwise there has been little shoal-water topographic change.

Geomorphic change on islands

Hurricane effects above high-water level on islands seem relatively permanent. Of the seven islands which disappeared in 1961, none has reappeared. In the intertidal and nearshore areas, small changes have occurred in beach profiles, with minor filling and cutting depending on exposure, but these changes are generally within the range of seasonal fluctuation. In the area of most intense reef damage slight erosion of seaward shores has occurred as a result of increased exposure. Above and away from the beaches there has been no topographic change, except at Northern Cay, Lighthouse Reef, where small dunes have formed under the lee of a hurricane-deposited sand sheet. Where vegetated islands were reduced to shifting sandbars by the storm, as at Goff's and Sergeant's Cays, continuous minor changes of form and location are taking place, typical of those of unvegetated islands. On islands where rubble and shingle bars and ridges were formed by the storm, these remain essentially intact, with slight changes in location and shape, and are now being vegetated. At Jaluit atoll, by contrast, where they were more extensive, such bars were largely eroded and pushed shorewards on the leeward sides of islands in the three years after the typhoon (Blumenstock *et al.*, 1961, p. 618). Gravel and sand sheets on islands and channels cut through island surfaces (except where the latter have been filled by man) remain essentially unaltered.

Many of the exposures of beachrock, cay sandstone and other rocks revealed by the hurricane, and then often poorly cemented, had been partially eroded by 1965, but the rocks which remained had been converted into tough limestones, and become relatively permanent features of the islands.

Vegetation change on islands

The major reef-island changes in the period 1962–5 have been vegetational. In the zone of major damage vegetation recovery, as at Jaluit, has depended largely on whether the original soil cover was preserved during the storm (mainly in the interiors of large islands), or whether it was either stripped by surface erosion or buried by fresh sand or rubble deposits (on the margins of larger islands, and across the whole surface of smaller islands near the storm centre). In the case of soil preservation, recovery of herbaceous vegetation and grasses has been rapid, and the ground cover of *Wedelia, Canavalia, Ipomoea* and *Stachytarpheta* may reach 100 per cent. Where soil stripping occurred, the vegetation cover is generally less than 25 per cent, and consists of discrete patches of *Sesuvium, Sporobolus, Ipomoea* and *Euphorbia*. In cases of soil burial, vegetation colonisation depends on the calibre and thickness of the hurricane deposits. Thick nearshore sand sheets may be so densely covered with *Ipomoea* and *Canavalia* that the physiographic effects of the hurricane are largely hidden. Deposits of shingle and rubble, especially where thin and overlying a deeply eroded cay surface, have a very patchy cover of *Euphorbia* and *Sesuvium*. Where whole islands were reduced to unvegetated sandbars by the storm, and in 1962 carried a pioneer strand flora dominated by *Portulacca oleracea*, the more stable areas are now covered with carpets of *Ipomoea, Canavalia, Euphorbia* and grasses.

The dominant nearshore shrubs destroyed by the hurricane in the zone of major damage have made a remarkable recovery. The stripped south shore of Half Moon Cay now has a 50 per cent cover of *Tournefortia gnaphalodes* in patches up to 30 ft in diameter and 4·5 ft tall. *Tournefortia* is the most widespread of the nearshore colonisers, but elsewhere *Suriana maritima, Sophora tomentosa, Borrichia arborescens* and *Conocarpus erecta* reach heights of 3–9 ft, regenerating from seed rather than stumps.

Tree recovery has been less spectacular. Coconuts which retained their crowns during the storm are now fruiting, and trees planted in 1962 now have heights of 3–9 ft. Fallen trunks and exposed root nets are rapidly rotting away. *Thrinax* has made a good recovery even where blanketed with shingle in nearshore areas. Of the littoral thicket dominants, *Cordia sebestena* has been most resilient, flowering freely in 1965 from the most shattered stumps; *Coccoloba uvifera*, while less widespread, has also regenerated from stumps in most cases; but *Bursera simaruba* appears to have been killed in all cases of severe damage.

While vegetation on the coral cays is rapidly recovering to normal, the defoliated mangrove suffered permanent damage; the 1965 survey showed that most of the defoliated mangroves in a 50-mile wide zone are dead. The boundaries of this zone are sharply delimited, and pass from mainly dead to mainly living mangrove within 5–10 miles. Within the mortality zone there are small patches of living *Rhizophora* in areas sheltered from extreme winds and waves, but otherwise vegetation growth is limited to fresh colonisation by *Batis maritima*, grasses, and sedges on higher land. The grey colour of the dead mangrove contrasts vividly from the air with the fresh green colour of drier, sandy areas with living vegetation. Similar contrasts have been noted after Hurricane 'Donna' in Florida (Craighead and Gilbert, 1962). As a result of this massive mortality, the formerly abundant *Rhizophora* seedlings are now rare in shallow water throughout Zone I, and the presence of the dead mangrove itself inhibits fresh growth.

Summary

After three years one can suggest, therefore, that many of the topographic changes in the islands are irreversible, against the time scale of the existence of the cays; and that the flora and fauna of the reefs and islands have a wide range of recovery periods. For the reefs this is likely to be measured in decades, and the sedimentary record of reef destruction of reef slopes will be permanent. For the vegetation the recovery period is least for quick-growing grasses, herbs and shrubs, many of which are rapidly colonising areas previously occupied by other vegetation or kept artificially cleared by man. The slower-growing vegetation responds less rapidly; it may be fifty years before coconuts and mangroves return to their pre-hurricane condition.

This analysis suggests that catastrophic storms, even of the magnitude of 'Hattie', represent only interruptions in a long-term equilibrium in reef and island geomorphology. The question arises as to whether any long-period trends can be recognised. The British Honduras data suggest they can (Stoddart, 1964).

IX. STORMS, MAN AND THE EQUILIBRIUM ISLANDS

The discussion of storm effects on vegetation in section VI demonstrated that on islands with natural woodland or dense coconut thicket fresh shingle is banked against the vegetation hedge, whereas on those cleared for coconuts, with little or no natural vegetation, islands are easily overtopped by waves and considerable erosion takes place. Half Moon Cay, on Lighthouse Reef, shows both kinds of effect: it is a sand-shingle island, 1,100 yds long, with shingle ridges along its south and north-east shores reaching a maximum height of 10 ft above sea-level (Fig. 9.12). The western half of the island in 1961 was covered with a thicket of *Cordia sebestena, Bursera simaruba, Ficus, Neea* and coconuts; the eastern half had been cleared of natural vegetation, mostly since 1927, and in 1961 carried tall coconuts, largely without ground vegetation. Before the hurricane the thicket-covered west end of the cay consisted of a seaward shingle ridge, declining in height from 9 ft to 3 ft above sea-level from east to west, with attendant decrease in shingle calibre; *Cordia* formed a dense hedge along the ridge crest. Lagoonwards of the crest the surface was flat, built of coarse coral grit and rotten shingle, and covered with dense vegetation. The eastern half of the cay, under coconuts, was almost entirely sand,

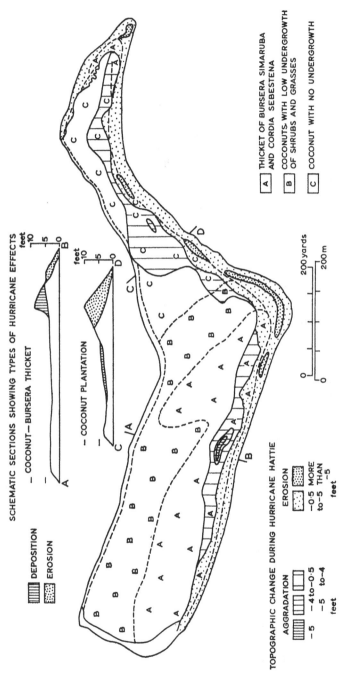

SCHEMATIC SECTIONS SHOWING TYPES OF HURRICANE EFFECTS

DEPOSITION
EROSION

— COCONUT—BURSERA THICKET

— COCONUT PLANTATION

TOPOGRAPHIC CHANGE DURING HURRICANE HATTIE

AGGRADATION
−5   −4 to −0·5
−5   to −4
feet

EROSION
−0·5 MORE
to −5   THAN
−5
feet

A   THICKET OF BURSERA SIMARUBA
    AND CORDIA SEBESTENA

B   COCONUTS WITH LOW UNDERGROWTH
    OF SHRUBS AND GRASSES

C   COCONUT WITH NO UNDERGROWTH

200 yards
200 m

Fig. 9.12   Topography change during Hurricane 'Hattie' at Half Moon Cay

the seaward beach crest rising usually to more than 8 ft, and locally to 10 ft above sea-level. During the hurricane, in addition to considerable wind damage to vegetation, changes resulting from south to south-west seas took the following forms:

(*a*) On the sand area under coconuts, the seaward beach retreated up to 20 yds, and the crest elevations were lowered at least 3 ft and in places up to 7 ft. The area back of the crest suffered surface sand stripping, with some channelling on the narrower part of the cay.

(*b*) On the western vegetated sector, the seaward ridge crest was pushed about 25 yds landwards and nearshore vegetation was uprooted and destroyed. The dense vegetation thicket, however, acting as a baffle against the waves, served as a massive sediment trap for coral blocks, shingle and sand. In place of the former graded sediment distribution and decline in crest height from east to west, fresh sediment has been piled against the vegetation barrier to a height of 8 ft above sea-level for most of its length, and in one place to 10 ft. The inner edge of the fresh sediment is marked by a sharp break of slope.

Thus, where the natural vegetation had been replaced by coconuts before the storm, erosion and breach retreat led to net vertical decreases in height of 3–7 ft; whereas where natural vegetation remained, banking of storm sediments against the vegetation hedge led to net vertical increases in height of 1–5 ft.

Similar changes were found throughout the reef area. Table I

TABLE I

Role of vegetation cover in hurricane effects on reef islands during Hurricane 'Hattie', 1961

Hurricane effect	*Vegetation type*					
	A	B	C	D	E	F
	Natural vegetation	Coconuts with regenerated thicket	Coconuts with low undergrowth	Coconuts with no undergrowth	Small island with small vegetation thicket	Un-vegetated
Disappearance	–	–	–	2	10	7
Major surface sand-stripping and channel-cutting	–	2	3	17	–	–
Major beach retreat	–	3	–	10	–	–
Marginal aggradation	15	9	–	–	–	–

Numbers refer to individual islands in each vegetation and damage class.

summarises types of geomorphic change for all islands studied, in terms of six categories of vegetation cover, and clearly demonstrates the importance of vegetation type in controlling the pattern of storm effects. Thus the climax vegetation of reef islands, with its dense structure and root systems, forms a natural protection against storm waves, and once such vegetation is established, it acts during violent storms as a sediment trap, leading to a net increase in height of the land surface. In 1961 storm sediments were thus built to a height of 10 ft above sea-level at Half Moon Cay, a height equal to that of any other island on the reefs before the storm. In the reef-island ecosystem, therefore, vegetation and sediment accumulation are interlocked; without the sediment-trapping function of the natural vegetation, islands could not be built to heights of more than 3–5 ft above sea-level unless there were relative shifts of sea-level. High-standing islands are thus in complex equilibrium with marine processes, with vegetation as a critical control during major storms. If the vegetation is removed from any island, this equilibrium is disturbed. Sediment is no longer trapped from storm waters, which cause considerable marginal and surficial erosion leading to net decrease in surface height and even to disappearance or fragmentation of the island. When natural vegetation is removed, it is normally replaced by coconut plantations, which (*a*) have an open structure easily penetrated by sea water; (*b*) frequently have no ground vegetation, thus exposing the surface to stripping and channelling; and (*c*) have a dense but shallow (12–18 in.) root mat easily undermined by marginal sand sapping. Hence in any given area subject to catastrophic storms, islands with natural vegetation will tend to increase in height during storm action and thus become progressively less subject to catastrophic damage, while those under coconuts will tend to be eroded and destroyed.

The many islands eroded by Hurricane 'Hattie' and the recent replacement of natural vegetation by coconuts, suggest that storm effects may have changed within historic times. In British Honduras coconuts are a post-Columbian introduction. They were first noted by Uring in 1720, but according to a sketch of Half Moon Cay in 1829 were still few in number. They apparently increased during the nineteenth century, and especially with the growth of the coconut export industry in the period 1850–1930. At Half Moon Cay most of the trees date from the 1920s. The vegetation change, therefore, has largely taken place within the last century. Apart from some eighteenth-century maps, the first detailed charts of the reefs date

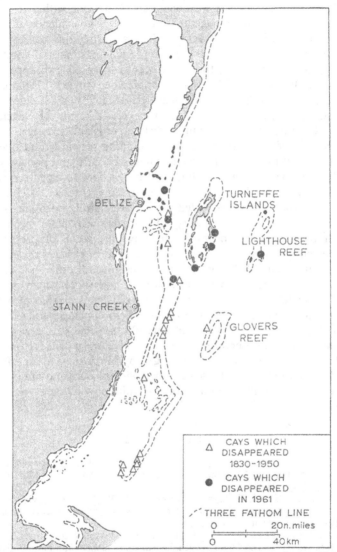

Fig. 9.13 Disappearance of cays on the British Honduras reefs since 1830

from the Admiralty surveys in the 1830s, since when approximately twenty islands have disappeared (Fig. 9.13). Many others have undergone erosion, especially on their seaward shores. Almost all islands described in 1760–1830 as being covered with bushes or

ACG G

thicket are now planted with coconuts, and those within the storm area suffered severe damage in 1961. There is no evidence of islands becoming established as vegetated cays during this same period. Concurrently with the widespread replacement of natural vegetation by coconuts, therefore, there has been a large-scale disappearance of formerly vegetated islands, in a period when (since 1787) there have been twenty-one recorded hurricanes, fourteen of them since 1900.

Island aggradation during storms, as a result of vegetation sediment trapping, will result in distinctive island structures and a succession of soil and root horizons dipping seawards and overlain by thick but narrow and variable bands of coarse and poorly sorted sediments. Trenching on Jaluit atoll has in fact revealed sequences of soil horizons buried by later sediments (McKee, 1959), and similar structures are reported from Ifaluk atoll, Caroline Islands (Tracey *et al.*, 1961). Storm aggradation also provides a mechanism whereby islands can be built up above the reach of present-day wave action, without the necessity of calling for changes in sea-level.

It thus seems likely that island structures will differ not only in those areas exposed to tropical storms, and those which are not, but also within the storm areas, on those islands cleared of natural vegetation and those which are not. These conclusions need to be tested by further field study in carefully selected reef areas, while at the same time observations must be continued in British Honduras, Jaluit atoll and other places damaged by hurricanes, in an effort to understand the role of these extremely violent but yet quite frequent tropical storms in moulding the morphology of coral reefs and islands.

REFERENCES

BAILEY, H. P. (1959) 'An analysis of coastal climates with particular reference to humid midlatitudes', *Louisiana State Univ. Coastal Studies Inst., 2nd Coastal Geog. Conf.*, 23–55.

BALL, M. M., SHINN, E. A., and STOCKMAN, K. W. (1963) 'Geologic record of hurricanes', paper presented at the Houston meeting of the Society of Economic Palaeontologists and Mineralogists.

BATES, F. L., FOGLEMAN, C. W., PARENTON, V. J., PITTMAN, R. H., and TRACEY, G. S. (1963) 'The social and psychological consequences of a natural disaster: a longitudinal study of Hurricane Audrey', *Nat. Acad. Sci. – Nat. Res. Council, Publication 1081*, 1–190.

BERGERON, T. (1954) 'The problem of tropical hurricanes', *Quart. J. Roy. Meteorol. Soc.*, LXXX 131–64.

BLUMENSTOCK, D. I. (1958a) 'The effect of typhoons upon the morphology of coral reefs and atolls (abstract)', *Ann. Assoc. Amer. Geogr.*, XLVIII 253–4.

—— (1958*b*) 'Typhoon effects at Jaluit atoll in the Marshall Islands', *Nature*, CLXXXII 1267–9.

—— (ed.) (1961) 'A report on typhoon effects upon Jaluit atoll', *Atoll Res. Bull.*, LXXV 1–105.

—— FOSBERG, F. R., and JOHNSTON, C. G. (1961) 'The re-survey of typhoon effects on Jaluit atoll in the Marshall Islands', *Nature*, CLXXXIX 618–20.

BOWMAN, H. H. M. (1918) 'Botanical ecology of the Dry Tortugas', *Papers from the Dep. of Marine Biol.*, *Carnegie Inst.*, XII 109–38.

BROWN, C. W. (1939) 'Hurricanes and shoreline changes in Rhode Island', *Geogr. Rev.*, XXIX 416–30.

CARTER, J. (1959) 'Mangrove succession and coastal change in south-west Malaya', *Trans. Inst. Brit. Geogr.*, XXVI 79–88.

CHAMBERLAIN, J. L. (1959) 'Influence of Hurricane Audrey on the coastal marsh of southwestern Louisiana', *Louisiana State Univ.*, *Coastal Studies Inst.*, *Tech. Rep. 10-B*, 1–38.

CRAIGHEAD, F. C. (1964) 'Land, mangroves and hurricanes', *Bull. Fairchild Tropical Garden*, XIX 1–28.

—— and GILBERT, V. C. (1962) 'The effects of Hurricane Donna on the vegetation of southern Florida', *Quart. J. Florida Acad. of Sci.*, XXV 1–28.

DAVIS, J. H. (1940) 'The ecology and geologic role of mangroves in Florida', *Papers of the Tortugas Lab.*, XXXII 302–412.

—— (1942) 'The ecology of the vegetation and topography of the sand keys of Florida', *Papers of the Tortugas Lab.* XXXIII 113–95.

DAVIS, W. M. (1914) 'The home study of coral reefs', *Bull. Amer. Geogr. Soc.*, XLVI 561–77, 641–54, 721–39.

DUNN, G. E., and MILLER, B. I. (1960) *Atlantic Hurricanes* (Baton Rouge: Louisiana State Univ. Press).

EMERY, K. O. (1962) 'Marine geology of Guam', *U.S. Geol. Surv.*, *Prof. Paper 403-B*, B1–B76.

ENGLE, J. B. (1948) 'Investigations of the oyster reefs of Mississippi, Louisiana and Alabama following the hurricane of September 19 1947', U.S. Fish and Wildlife Service, Special Scientific Report, *Fisheries*, LIX 1–70.

FINCKH, A. E. (1904) *Biology of the Reef-forming Organisms of Funafuti Atoll. The Atoll of Funafuti: Borings into a Coral Reef and their Results; being the Report of the Coral Reef Committee of the Royal Society*, 125–50.

GLEGHORN, R. J. (1947) 'Cyclone damage on the Great Barrier Reef', *Rep. Gt. Barrier Reef Cttee.*, VI (1) 17–19.

GLYNN, P. W., ALMODÓVAR, L. R., and GONZÁLEZ, J. G. (1965) 'Effects of Hurricane Edith on marine life in La Parguera, Puerto Rico', *Caribbean J. Sci.*, IV 335–45.

GOODBODY, I. (1961) 'Mass mortality of a marine fauna following tropical rain', *Ecology*, XLII 150–5.

GOREAU, T. F. (1964) 'Mass expulsion of zooxanthellae from Jamaican reef communities after Hurricane Flora', *Science*, CXLV 383–6.

HAYES, M. O. (1965) 'Sedimentation on a semiarid, wave-dominated coast (South Texas); with emphasis on hurricane effects', Ph.D. thesis (University of Texas, Austin) 1–350.

—— (1966) 'Some observations on the geological effects of hurricanes, south Texas coast', *Bull. Houston Geol. Soc.* IX 18–26.

HEDLEY, C. (1925) 'The natural destruction of a coral reef', *Rep. Gt. Barrier Reef Cttee.*, I, 35–40.

HOWARD, A. D. (1939) 'Hurricane modification of the offshore bar of Long Island, New York', *Geogr. Rev.*, XXIX 400–15.

JOUBIN, L. (1912) 'Carte des bancs et réclifs de coraux', *Ann. de l'Inst. Océanogr.*, *Monaco*, IV 1–7, 5 maps.

LESSA, W. A. (1964) 'The social effects of Typhoon Ophelia (1960) on Ulithi', *Micronesica*, I 1–47.

MCGILL, J. T. (1958) 'Map of coastal landforms of the world', *Geogr. Rev.*, XLVIII 402–5.

MCINTIRE, W. G., and WALKER, H. J. (1964) 'Tropical cyclones and coastal morphology in Mauritius', *Ann. Assoc. Amer. Geogr.*, LIV 582–96.

MCKEE, E. D. (1959) 'Storm sediments on a Pacific atoll', *J. Sedimentary Petrology*, XXIX 354–64.

MOORHOUSE, F. W. (1936) 'The cyclone of 1934 and its effects on Low Isles, with special observations on *Porites*', *Rep. Gt. Barrier Reef Cttee.*, IV 37–44.

MORGAN, J. P. (1959) 'Coastal morphological changes resulting from Hurricane Audrey', *Proc. Salt Marsh Conf.*, *Sapela I.*, *Georgia, 1958* (Athens, Georgia) 32–3; discussion, 33–6.

—— NICHOLS, L. G., and WRIGHT, M. (1958) 'Morphological effects of Hurricane Audrey on the Louisiana coast', *Louisiana State Univ.*, *Coastal Studies Inst.*, *Tech. Rep. 10-A*, 1–53.

NEWELL, N. D. (1961) 'Recent terraces of tropical limestone shores', *Zeitschrift für Geomorphologie*, n.f., Supplementband 3 (R. J. Russell (ed.), *Pacific Island Terraces – Eustatic?*) 87–106.

NICHOLS, R. L., and MARSTON, A. C. (1939) 'Shoreline changes in Rhode Island produced by hurricane of September 21, 1938', *Bull. Geol. Soc. Amer.*, L 1357–70.

OPPENHEIMER, C. H. (1963) 'Effects of Hurricane Carla on the ecology of Redfish Bay, Texas', *Bull. Marine Sci.*, *Gulf and Caribbean*, XIII 59–72.

PUTNAM, W. C., AXELROD, D. I., BAILEY, H. P., and MCGILL, J. T. (1960) *Natural Coastal Environments of the World* (Los Angeles: Univ. of California) 1–140.

SAUER, J. D. (1959) 'Coastal pioneer plants of the Caribbean and Gulf of Mexico' (Univ. of Wisconsin, Depts. of Botany and Geography, mimeographed).

—— (1962) 'Effects of recent tropical cyclones on the coastal vegetation of Mauritius', *J. Ecology*, L 275–90.

STEERS, J. A. (1937) 'The coral islands and associated features of the Great Barrier Reefs', *Geogr. J.*, LXXXIX 1–28, 119–46.

—— (1940) 'The coral cays of Jamaica', *Geogr. J.*, XCV 30–42.

—— Chapman, V. J., COLMAN, J., and LOFTHOUSE, J. A. (1940) 'Sand cays and mangroves in Jamaica', *Geogr. J.*, XCVI 305–28.

STEPHENSON, W., ENDEAN, R., and BENNETT, I. (1958) 'An ecological survey of the marine fauna of Low Isles, Queensland', *Autral. J. Marine and Freshwater Res.*, IX 261–318.

STODDART, D. R. (1962) 'Three Caribbean atolls: Turneffe Islands, Lighthouse Reef, and Glover's Reef, British Honduras', *Atoll Res. Bull.*, LXXXVII 1–151.

—— (1963) 'Effects of Hurricane Hattie on the British Honduras reefs and cays, October 30–31, 1961', *Atoll Res. Bull.*, XCV 1–142.

—— (1964) 'Storm conditions and vegetation in equilibrium of reef islands', *Proc. 9th Conf. Coastal Eng.* (Lisbon) 893–906.

—— (1965) 'Re-survey of hurricane effects on the British Honduras reefs and cays', *Nature*, CCVII 589–92.

—— (1968a) 'Reconnaissance geomorphology of Rangiroa atoll, Tuamotu Archipelago', *Atoll Res. Bull.*

——(1968b) 'Post-hurricane changes on the British Honduras reefs and cays: re-survey of 1965', *Atoll Res. Bull.*

TABB, D. C., and JONES, A. C. (1962) 'Effects of Hurricane Donna on the aquatic fauna of North Florida Bay', *Trans. Amer. Fisheries Soc.*, XCI 375–8.

TANNER, W. F. (1961) 'Mainland beach changes due to Hurricane Donna', *J. Geophys. Res.*, LXVI 2265–6.

THOM, B. G. (1967) 'Mangrove ecology and deltaic geomorphology: Tabasco, Mexico', *J. Ecol.*, LV 301–44.

THOMAS, L. P., MOORE, D. R., and WORK, D. C. (1961) 'Effects of Hurricane Donna on the turtle grass beds of Biscayne Bay, Florida', *Bull. Marine Sci., Gulf and Caribbean*, XI 191–7.

TRACEY, J. I., ABBOTT, D. P., and ARNOW, T. (1961) 'Natural history of Ifaluk atoll: physical environment', *Bishop Museum Bull.*, CCXXII 1–75.

U.S. ARMY ENGINEER DISTRICT (1962) *Report on Hurricane Carla, 9–12 September 1961* (Galveston, Texas: U.S. Army Engineer District, Corps of Engineers).

VERMEER, D. E. (1963) 'Effects of Hurricane Hattie, 1961, on the cays of British Honduras', *Zeitschrift für Geomorphologie*, n.f., VII 332–54.

WARNKE, D. A. (1967) 'Conditions of beach retrogression in a low-energy environment', *Zeitschrift für Geomorphologie*, n.f., XI 47–61.

—— GOLDSMITH, V., GROSE, P., and HOLT, J. J. (1966) 'Drastic beach changes in a low-energy environment caused by Hurricane Betsy', *J. Geophys. Res.*, LXXI 2013–16.

WELLS, J. W. (1957) 'Coral reefs', *Geol. Soc. Amer. Memoirs*, LXVII (*Treatise on Marine Ecology and Palaeoecology*, I) 609–31.

YAMASHITA, A. C. (1965) 'Attitudes and reactions to Typhoon Karen (1962) in Guam', *Micronesica*, II 15–23.

10 The East Coast Floods 31 January–1 February 1953

J. A. STEERS

THE FLOODING which took place at so many places on the east coast between the Tees and Dover on the night of 31 January 1953 was serious both to life as well as to property, whether houses or land (Fig. 10.1). It is estimated that 206,161 acres[1] were definitely flooded (and possibly another 11,732), and that 307 people lost their lives. It is impossible to minimise this disaster, but nevertheless it is right to realise that the results might have been far more serious. We are more than fortunate in that the rivers, particularly those flowing through the fenlands, were not in flood; we were fortunate too in that the surge did not occur at the top of the tide and, what is of still greater significance, did not occur on a high spring tide (Bowden, 1953). The predicted tides for 31 January were 1–3 ft less – according to locality – than can occur at other times of the year. There is little doubt that if the surge had occurred under conditions of a high spring tide and flooded rivers, the whole of the Fens, and other similar places, would have been submerged, mainly by fresh water.

Surges are not uncommon, but it is only in comparatively recent years that they have been investigated. It is a matter of great regret that Dr R. H. Corkan, who had studied the subject more fully than anyone else in this country, died a young man, only a few years ago. The account he gave of the surge on 8 January 1949, and of the conditions under which it was generated, is applicable in a remarkable way to the surge of 31 January 1953. On the other hand the violence in 1953 of the storm in the northern part of the North Sea was exceptional. The devastation of the forests in north-eastern Scotland, moreover, and the abnormally high northerly winds in the Shetland–Orkney area meant that, despite the far less disastrous wind action in the southern part of the North Sea, there was very rough water – far more so than is usually associated with such north-westerly, or north-north-westerly winds, as prevailed.

[1] I am particularly indebted to Dr E. C. Willatts and his colleagues in the Ministry of Housing and Local Government, who gave great help with the maps. The figures are from measurements made in Dr Willatts's office.

Fig. 10.1 Flooded areas on the east coast, 1953

In this paper no account is given of the profound damage that occurred in Holland. The surge which travelled southwards along our east coast later turned and moved northwards along the coast of the Low Countries, where it was accompanied by an onshore wind in which occasional gusts exceeded 90 m.p.h. (Ufford, 1953). It needs but little imagination to understand what happened under such circumstances to a country almost entirely protected by dykes or dunes.[1]

THE METEOROLOGICAL CONDITIONS

It is interesting to consider the sequence of events from noon on Thursday, 29 January, until noon on Sunday, 1 February. Fig. 10.2 shows the position of the centre of the depression for every six hours during that time. It will be noted that the pressure at the centre of the depression fell from 1004 millibars at noon on 29 January, to less than 968 mb. at 6 a.m. and noon on 31 January. During this period the winds strengthened, so that the forecast issued at noon on Saturday, 31 January said 'All districts will have gale-force winds, severe in many places, and squally showers, mainly of hail or snow. Considerable snowfall may occur over high ground. Thunderstorms will occur here and there. It will be cold.'[2]

The general track of the depression from the Atlantic, around the north of Scotland into the North Sea, and the accompanying movement of the anti-cyclone behind it meant that a very powerful northerly airstream swept down the eastern side of Britain and the western part of the North Sea. The records of the Daily Weather Reports taken individually do not perhaps suggest excessive wind speeds: at midnight, 31 January (i.e. Friday–Saturday night) 50 knots were recorded for Benbecula and Stornoway; at 6 a.m. on Saturday Benbecula recorded 53 knots, and over the north of Scotland wind speeds between 35 and 45 knots were common. Farther south, along the east coast, London Airport recorded 22, Felixstowe 28, Gorleston 18, Spurn Head 25, Tynemouth 24 and Leuchars 26. At midnight, Saturday–Sunday, the wind speed at London Airport was given as 13 knots, 34 knots at Gorleston and 16 at Leuchars.

[1] I was able to visit most of the flooded areas very soon after the disaster, thanks to the Nature Conservancy, who allowed me to extend a research grant for the purpose.

[2] In this section I have made great use of the Daily Weather Reports of the Meteorological Office.

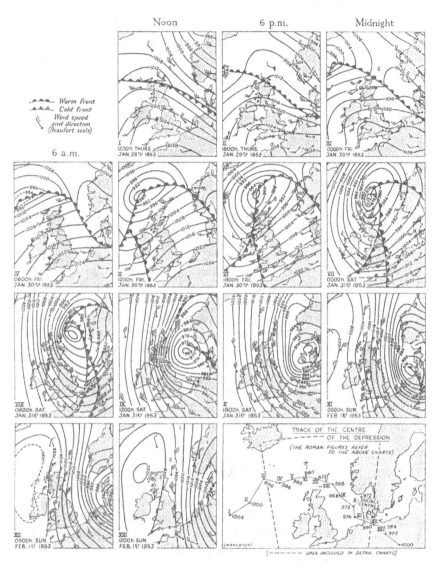

Fig. 10.2   Synoptic charts for every six hours from noon on 29 January to noon on 1 February 1953 (based on daily weather reports of the Meteorological Office)

ACG H

These speeds, however, are somewhat misleading if taken by themselves. The great damage to forests in northeastern Scotland in itself implies something quite exceptional. The wind record at Aberdeen University station (to some extent sheltered) shows gusts up to 75 knots, and from about 9 a.m. to 9 p.m. on 31 January the velocities were high, many gusts throughout the whole day exceeding or reaching 50 knots. Velocities of hurricane force are recorded on the chart for midday of 31 January. The wind velocities shown in Fig. 10.3 give a better impression of the severity and length of the storm.[1]

Along the east Scottish coast, and probably as far south as Flamborough Head, the winds were abnormally strong, although they abated somewhat farther south on our coasts. The Low Countries on the other hand suffered almost the full force of the storm.

What is of great significance in relation to sea conditions is the sequence of winds in the North Sea basin. At 6 a.m. on Thursday, 29 January the winds in the North Sea were from a little north of west on the English side, becoming more westerly on the Dutch side. Much the same conditions prevailed throughout Friday except that there was rather more south in the winds in the Flemish Bight. At 6 p.m. on Friday the winds over nearly all the North Sea were approximately south-west, and by midnight this was accentuated, so that at Felixstowe the velocity was 18 knots from a direction 20° west of south and at Gorleston 23 knots from 10° west of south. On the coast of Belgium and Holland there were fresh to strong breezes from south-west to south-south-west.

As the depression advanced into the North Sea it became more intense, and the maximum gradient behind the centre held from midnight to 6 a.m. on 1 February. Thus both maximum depth and gradient occurred when the depression was in the North Sea. At noon on Saturday winds of hurricane force are indicated on the chart between Orkney and Moray. These were blowing from almost due north. All along the east coast of Scotland velocities of 50 or more knots are recorded, the direction being in general about north-west. At 6 p.m. there was still a fresh gale, the winds being onshore on the south side of the Moray Firth and at Hunstanton, but otherwise obliquely offshore from north-north-west except for parts of north Norfolk. It was the worst northerly gale on record in the British Isles.

[1] 'On Costa Hill, Orkney, several gusts, exceeded 105 knots (121 m.p.h.); the highest reached 109 knots' (M.W. Report, Jan 1953).

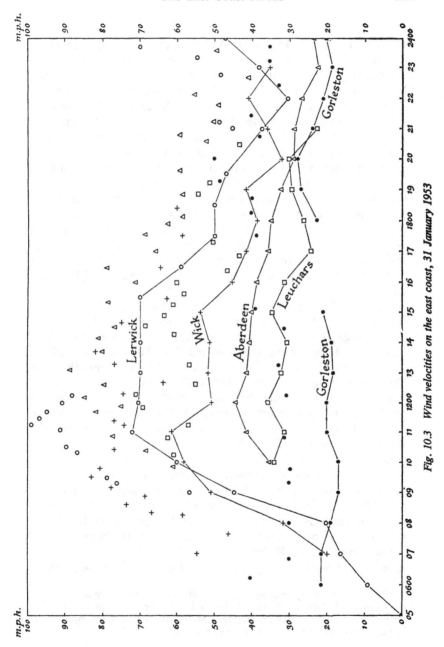

Fig. 10.3 Wind velocities on the east coast, 31 January 1953

TIDAL CONDITIONS

Full moon was at 11.44 p.m. at Cambridge on Thursday, 29 January. At Hull and Leith the maximum predicted tide of that particular set of springs fell in the afternoon of Sunday, 1 February. At London Bridge the maximum was predicted for the morning tide of Sunday. None of these maxima, however, was noticeably a high one.[1] At London Bridge the predicted tide may be rather more than 2 ft higher than that on 31 January, that at Hull may be 3 ft higher and at Leith rather more than 2 ft. Thus, as far as ordinary tidal maxima were concerned, there was no reason to expect anything unusual.

Where, however, the actual tides are examined, it will be apparent that there was a great discrepancy between them and the predicted tides. The curves in Fig. 10.4 show that at Aberdeen the day-time tide on 31 January was nearly 3 ft, that at Immingham 7 ft, that at Southend nearly 8 ft and that at Dover approximately 6 ft higher than predicted. Where records are available it will also be noticed that the height of the next succeeding tide was rather higher than anticipated.[2]

The third line on Fig. 10.4 is obtained by taking the difference between the predicted and actual curves – and it is thus the measure of the surge. The maximum of the surge was usually about three hours later than the maximum of the tide. The amplitude of the surge amounted to about 2 ft at Aberdeen, 7 ft at the mouth of the Tees, 7½ at Immingham, about 9 at Southend, and about 7 at Dover.

The surge swept along the east coast of Britain from north to south and sent off branches into inlets like the Wash and the Thames estuary. It may in some ways be compared to a tidal bore. The *Admiralty Manual of Tides* (1941) defines a surge as 'a water movement which is quickly generated and whose effects are soon over' and then continues as follows: 'The rapidity of generation, the notable rise (or fall) of the sea level, and the manner in which the surge travels, are characteristics which distinguish it from other meteorological disturbances of sea level. Clearly the mechanism of generation and propagation is not unlike that of the bore. . . .'

Because of the small size of the North Sea the local tides result from oscillations travelling in from the Atlantic round the north of

---

[1] The eclipse on the Thursday–Friday night had no appreciable effect.
[2] I am much indebted to the Hydrographer to the Navy, to Commander Farquharson, R.N., and to my colleague Mr W. W. Williams, for information in this section, and also for the curves in Fig. 10.4.

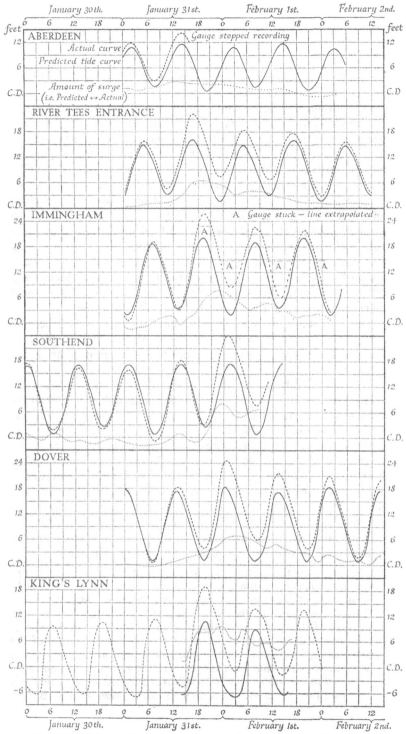

**Fig. 10.4 Actual and predicted tidal curves on the east coast, 30 January–
2 February 1953**

Scotland. Because of the shape, size and depth of the North Sea three nodal lines occur (see Fig. 10.5). Between any two lines, high (or low) water will occur simultaneously, and the maximum range will be half-way between any pair of lines, and in crossing any line one passes from an area of high (or low) water to one of low (or high) water. If now the effects of gyration are added, forces are brought into being which deflect the tidal currents to the right and so at certain hours gradients occur along the nodal lines. In other words there will be oscillations along the nodal lines, but at some point on any given line the water-level will remain constant. This is the amphidromic point, around which the tidal oscillations turn in an anti-clockwise direction and produce the conditions shown in Fig. 10.5. One effect of friction is a loss of energy which in turn has a diminishing effect on the range of the tides and also produces a difference between the eastern and western shores. This is most markedly the case in the northern part of the North Sea.

Fig. 10.6 makes it clear that the high (or low) waters become progressively later from north to south along the east coast of Britain. The surge followed this direction, and it is clear that a warning system can, in principle, be quite effective, since the surge arrives considerably later off, say, East Anglia, than it does off Dunbar and because the height of the surge may also increase in the same direction. At King's Lynn, for example, the predicted tide was 22.9 ft above the dock sill; the actual tide reached 31 ft – the surge, in other words, was 8.1 ft high. This, in the opinion of Mr Doran, Chief Engineer of the Great Ouse River Board, was the greatest that had been experienced at Lynn for many years, possibly for centuries.

In considering the tidal and surge phenomena which accompanied this storm, it is interesting to refer to the North Sea levels during the storm of 8 January 1949. The meteorological conditions on that occasion and on 31 January 1953 had many points in common – so much so that it is relevant to quote from Corkan's paper (1950):

> The origin of the disturbance as indicated by the weather maps was a deep depression which developed very rapidly off the north-west coast of Scotland and then passed quickly eastwards across the extreme north of the North Sea and southern Norway. The passage of the centre of the depression across the North Sea was accompanied by a rapid veering of the winds in the rear of the depression, from south-west to north-west and north, and it is of some importance that for a short time strong northerly winds were localised in the western half of the North Sea and were particularly affecting the east coast of the British Isles. These conditions were

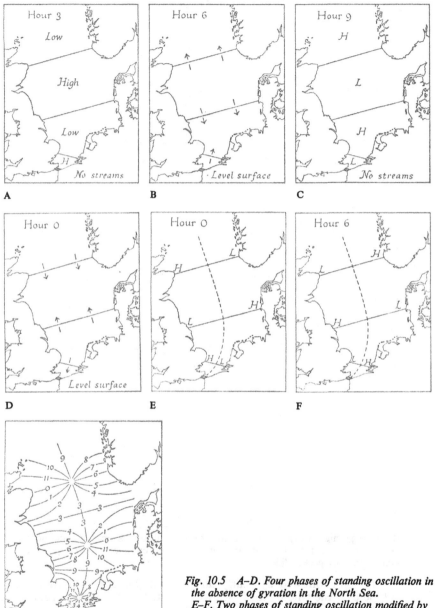

Fig. 10.5 A–D. Four phases of standing oscillation in the absence of gyration in the North Sea.
E–F. Two phases of standing oscillation modified by gyration. G. Co-tidal lines deduced from particular phases of standing oscillation, modified by gyration.

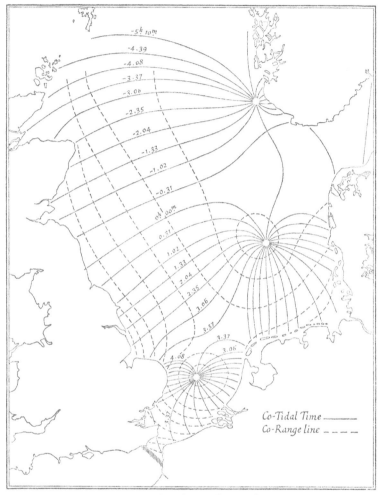

Fig. 10.6 The North Sea tides
On average, high water occurs simultaneously at all points on any continuous line, in time order from the standard port (river Tyne entrance). The range is the same along any pecked line.

very soon followed by nearly uniform conditions of a northerly type over the whole North Sea which lasted until the disturbance subsided.

In this case also the surge travelled down the east coast, and at King's Lynn it reached 5·5 ft high, which was its maximum. This disturbance took ten hours to travel from Dunbar to Southend, and Corkan found that the speed of transmission of a surge and that of the diurnal tide along our coast are almost identical. The tidal conditions were not, however, quite comparable to those on 31 January of this year, since on 7 January 1949, the day before the surge, the moon was in its first quarter, and so neap tides prevailed.

THE EFFECTS OF THE FLOODING ON THE COASTAL AREAS

Some damage occurred in Northumberland, Durham and on the coasts of north-east Yorkshire, but in Holderness, near Easington, the sea broke over the dunes and the flood water made exits to the Humber near Skeffling. Spurn Head does not appear to have suffered much change.

The Lincolnshire coast is nearly all low and flat. There is a small cliff at Cleethorpes, now protected by the promenade. From there right round to the Wash no solid ground reaches the sea. Nearly all the area – known as Lincoln Marsh – is reclaimed ground protected by walls and dunes. Between Cleethorpes and Donna Nook there was some flooding, and breaches were made in walls. The airfield at Northcoates Point was submerged. Southwards from Donna Nook, even as far as the northern end of Mablethorpe, the amount of damage was relatively small. Where the dune belt and the beach are both wide the sea did little harm. Part of the energy of the waves was expended before they reached the dunes. These were certainly cut away, often deeply, but they made a real barrier. The most striking effect of dune protection was to be seen at Mablethorpe. About a mile north of the Coastguard Station the dunes held, even though they are much narrower than at Theddlethorpe. But in Mablethorpe itself there were serious flooding and damage. This was continuous with the damage at Trusthorpe and Sutton, and represented the most impressive on the whole of the east coast. Between Mablethorpe and Sutton the dunes were narrow, for the most part protected by sea walls. The sea-level rose higher than the pull-overs to the beach and flooded the land within. The narrow beach allowed access of bigger waves, and the abnormal height of the water washed away dunes and walls and invaded the towns. The damage in this area was considerable, and once the dunes or wall had disappeared the sea in some places could directly attack the houses. This was so at the southern end of Sutton where the

ACG H 2

Beach Hotel suffered. It must be remembered, too, that in Mablethorpe and Sutton, and to a lesser extent in Trusthorpe, the main
coast road runs immediately inside the dunes or wall and is particularly vulnerable.

South of Sutton the 'Roman' Bank acted as a very sound inner
bank and was not breached. At Anderby the dunes are somewhat
wider, and although seriously eroded, the line remained fairly
intact. Several houses on the dunes were partially undermined.
Much the same occurred at Chapel Point, where the sea swept over
the wall. The dunes to the north, although not very wide, stood up
well. Serious breaches were made at Ingoldmells, and there was a
good deal of flooding – including the large Butlin's Camp – south of
the point. With the widening of the beach towards Skegness the
damage decreased, and no marked alteration took place at that
town or at Gibraltar Point.

Miss King and Mr Barnes, both lecturers in geography at the
University of Nottingham, had studied Gibraltar Point in some
detail, and after the flood examined not only it, but also took
profiles through the beach and dunes along nearly all the Lincolnshire coast (King and Barnes, 1953). Their sections are shown in
Fig. 10.7 and illustrate remarkably clearly the effects of steeper and
narrower foreshores in leading to greater destruction of dunes and
sea walls. One effect of the storm was to scour away the beach and
leave extensive areas of mud, and in some places peat, exposed.
Much of this beach material was thrown inland.

Apart from occasional flooding, on account of high levels, little
happened in most parts of the Wash. The surge at King's Lynn was
estimated at about 8·1 ft. Most fortunately the Fen rivers were not
in flood. All the sea banks in the territory of the Great Ouse River
Board have been raised and strengthened in the last half century.
The reason for the breach in the Ouse walls at Wiggenhall was
primarily the unusual height of the surge, which overtopped the
wall. The river is tidal up to Brownshill Staunch. The high-water
level at Lynn was 18–21 in. above the level of the sea at Hunstanton,
and the level at Denver was somewhat below that at Lynn. It was
too much for some of the banks. The cause of the bank failures was
erosion of the *back* of the bank, except possibly in one case where
it may have been due to a rat hole.[1]

On the east side of the Wash at Heacham and Hunstanton there
was much damage and loss of life. The bungalows and other light

[1] Information in letter from Mr Doran (author's italics).

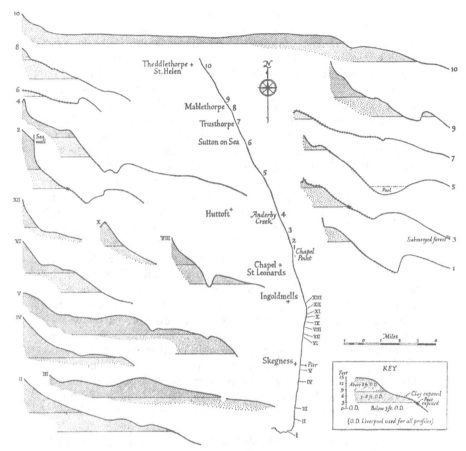

Fig. 10.7 Profiles along the Lincolnshire coast (King and Barnes, 1953)

buildings faced west, and even the narrow width of the Wash seems to have had some effect, since the wind and waves were directly onshore. Moreover, behind most of the bungalows the ground-level falls for a short distance, so that it was impossible for people to make quickly for higher ground. The damage to the promenade at Hunstanton was noteworthy, especially in view of its relatively sheltered position. The cliffs were also cut away here and there.

Along the coast of north Norfolk, from Old Hunstanton to Weybourne, the most obvious effect of the storm was the flooding of all the reclaimed marshes. The sea walls were overtopped, and breaches formed in most cases from the rear. The great bank running from Wells Quay to the harbour mouth was partly breached in this way, and walls and outhouses facing the marshes were knocked down by the waves inside this wall. The water ran up the old Warham Slade and for a time was 3 or 4 ft deep across the main road at the head of the Slade. The water from the entrance to the Slade also swept into the town and for a time submerged the station.

On the other hand natural salt marshes suffered not at all. Dunes were often eroded, but they too often stood remarkably well. On Scolt Head Island the seaward foot of the high dunes near the Hut was cut back 8–10 yds, and changes of that order, or even more, occurred in lines of dunes facing west or north-west. Low dunes were overrun and often washed away. Shingle spits were driven in, but apart from the washing away of the low dunes which face much of the foreshore of the island, the waves did little else than over-run the beach, somewhat flatten it and wash shingle and sand on to the marshes. Only in one place, already partially breached, does it seem likely that a long-lived gap will endure. Blakeney Point suffered more, partly because along much of it there is deeper water. The Headland and Far Point, in many ways like the corresponding features at Scolt Head Island, were cut into and partly washed away. The main shingle beach was overswept and pushed inland throughout its length. Opposite Salthouse a definite breach was made. This village suffered greatly, and not for the first time in its long history. Since there is high ground immediately behind, refuge was easy. At Cley and Wiveton flooding was severe, and conditions in the Glaven Valley must have resembled those in the great flood of 1897.

Along the cliffed coast of Norfolk there was some loss, particularly where the cliffs are low as at Bacton and Walcott. The main street of Walcott was destroyed, and many buildings wrecked. At Shering-ham the cliffs behind the promenade were severely attacked; Cromer, too, suffered somewhat in this way. Erosion has long been serious hereabouts, and throughout their length the cliffs offer very little resistance. Moreover, the beach tends to steepen, especially beyond Happisburgh. A few miles farther east the cliffs cease and give place to a dune coast. The dunes were 'trained' and blow-outs mended, the summit of the sandhills was kept fairly level, and marram grass planted. Nevertheless this is a very vulnerable part of the coast, and in 1938 there was a serious break at Horsey which led to the flooding of 7,469 acres. In January 1953 the only break occurred at Palling, but the dunes were seriously pared away all along this coast, and it was more than fortunate that storm and surge conditions were relatively short-lived, otherwise severe flooding in Broadland would have taken place.[1] The wall built across the 1938 gap at Horsey helped greatly in maintaining this stretch. At Winterton the dunes widen again, but suffered loss all along their front. Here too the beach is fairly steep.

In general, eastwards of Cromer the winds were more and more offshore, and beyond Winterton Ness and as far as the Thames estuary this was especially noticeable. In terms of direct erosion the sea did comparatively little damage except locally. At Kessingland, for example, at the northern and unfinished end of the new sea wall the cliffs were attacked at an unusually high level and were cut back. The same is true of the soft cliffs at Covehithe. At Dunwich the cliffs, equally soft, scarcely suffered at all at their northern end. On the other hand flooding was often serious, and large areas of Lowestoft were covered with water which came in both behind and over the sea walls, which themselves stood rigid. Even as far inland as Haddiscoe the water coming in through Lowestoft, and to some extent through Yarmouth, caused no little damage to a railway embankment. All the valleys, many of them blocked by shingle bars, along the Suffolk coast were flooded. At the south ends of Southwold and Aldeburgh the low-lying ground was inundated and damage done to property. The bungalows that suffered at Southwold were situated behind a sea wall on and in front of which small dunes had formed, and were in much the same position as were the houses at Mablethorpe and Sutton. But the barrier between them and the sea was much weaker.

The Alde did not break through at Slaughden. The walls on its north bank broke, and the shingle bank between Aldeburgh and Slaughden was flattened. For a time there was an overflowing of the sea at high water, but the river maintained its course. Even if matters had been left to nature I think this would have been so, but the flattened shingle was bulldozed up from the inside, and a new, even if thinner, ridge was soon made. However, the original ridge all the way from Aldeburgh to and beyond the Martello tower was displaced perhaps 50 yds landwards.

Apart from flooding, which occurred all along the coast, the breaking of some dykes and local erosion of cliffs, there was nothing of major importance until, just to the south of Clacton, the bungalow

[1] The width of the dune belt (1953) near Waxham is about half what it was.

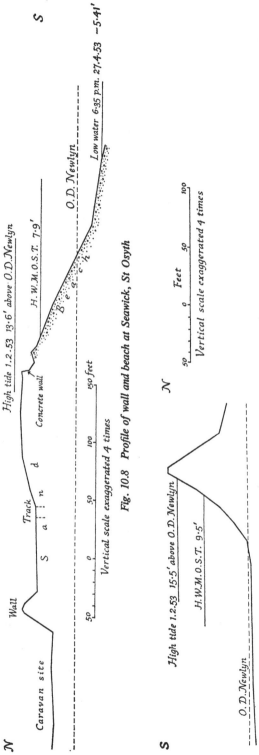

N

Caravan site

Wall

S   Track
a   n   d

Concrete wall

High tide 1.2.53 13·6' above O.D.Newlyn

H.W.M.O.S.T. 7·9'

O.D. Newlyn

Low water 6·35 p.m. 27.4.53 −5·41'

S

B
e   d   c   b

0   50   100   150 feet

Vertical scale exaggerated 4 times

Fig. 10.8 Profile of wall and beach at Seawick, St Osyth

N

High tide 1.2.53 15·5' above O.D.Newlyn

H.W.M.O.S.T. 9·5'

O.D.Newlyn

S

50   0   50   100
Feet

Vertical scale exaggerated 4 times

Fig. 10.9 Profile of sea wall near Kynoch Hotel, Canvey Island

town of Jaywick was severely hit. The sea once again overtopped the wall at the south end of the settlement and soon made a breach. The water also came in from behind. The whole place is below high-water mark, and flooding to a depth of 5 ft or so occurred over much of it. The caravans, as at Ingoldmells and other places in Lincolnshire, were swept far from their original sites and usually much broken. Even some permanent bungalows floated away and overturned. But the masonry sea wall along the main front was undamaged (Fig. 10.8).

The same kind of calamity affected Canvey Island. The major walls along the Thames held, and the flooding took place mainly over the lower and weaker walls along Benfleet and other creeks. Here too, once the walls were overtopped, these were eroded from the inside. Canvey has far more permanent inhabitants than Jaywick, and so material damage and loss of life were very serious. In both cases the settlements are below high-water mark. At Foulness Island the walls at the north and north-east were breached, and another big gap was made opposite Wakering. The whole island, together with its smaller neighbours, was inundated, and much valuable arable land spoiled (Fig. 10.9).

On the south side of the Thames estuary the winds were more directly onshore. There was considerable erosion in Sheppey and also in the soft cliffs at Beltinge just east of Herne Bay. The ruined church of St Mary at Reculver had been protected, but there was some erosion there. The low ground west of Whitstable was flooded, and huts and houses at Tankerton damaged. The low ground between Reculver and Birchington was flooded, and the main north Kent railway put out of action. Beyond the North Foreland there was far less damage, and although flooding occurred, the effects of the storm were of minor significance.

In the Thames and other rivers as far north as the Tees, inundations caused by overtopping of walls and the formation of breaches took place. There is no need to enumerate all these. The same series of events occurred in each place, and Fig. 10.1 shows far better than a long description how much of the coastal district was flooded.

EFFECT ON INDUSTRY

The damage to industry is perhaps not so fully realised as is that to domestic housing. The flooding in rivers and estuaries was extremely serious in this respect, and damage including that to the West

Norfolk Farmers Manure Co. at King's Lynn and to Birds Eye Foods Ltd and other firms at Yarmouth was great. In the Orwell estuary there was also severe flooding; twenty-two firms, as well as a gasworks and a power station, were seriously affected. Much the same happened in the river Colne, and some works here and elsewhere had to cease production entirely for periods of one to four weeks. The same conditions, often intensified, occurred in the Thames, and factories at Purfleet and other places, including those of Messrs Van den Bergh and Jurgens, and the Thames Board Mills Ltd, are examples of many that suffered seriously.

In estimating the effects of the storm, full allowance must be made for the loss suffered by industry both as a direct result of the flooding and also because of the fall in production as a consequence of it. Many industrial plants are situated on low ground and so occupy land which is often better used for other purposes. Moreover, proximity to navigable water is essential, and adequate protection of works and factories in these situations is of vital importance to the country as a whole.

Transport was locally much disorganised; 220 miles of track and eighty stations were temporarily out of service. Many hundreds of railway carriages were flooded. Several premises of London Transport, British Road Services and the Tilling Group were also flooded. Docks at King's Lynn, the Humber ports and the Hartlepools suffered, as also did stocks in the warehouses (*British Transport Review*, 1953; *Railway Magazine*, 1953).

THE EFFECTS ON AGRICULTURAL LAND

Land flooded by salt water presents several problems. Flooding, whether by salt or fresh water, temporarily destroys the tilth and makes the soil pasty and largely impervious to air. Salt has the further disadvantage of causing damage to crops, the extent of which will vary with the salt tolerance of the crop – barley, for example, can tolerate up to 0·25 per cent and some grass up to as much as 0·3 or 0·5 per cent. Evaporation in summer may raise the salt content to lethal proportions.

But the more serious long-term damage arises as a result of the substitution of calcium by sodium in the clay fraction of soils. This is a factor which is of the greatest significance in heavy soils, which may become very sticky and dry into hard lumps which will not crumble when subjected to frost or wetting.

It is essential to get rid of the flood water as soon as possible by whatever form of draining is most appropriate. Sometimes flooding with fresh water may be proper, and the water should be applied as fast as the soil can absorb it. This process is unlikely to meet with much success on heavy land.

There is no simple method of estimating how long land will be out of production, because there are too many variable factors such as the permeability of the soil, the efficiency of the drainage system, the local *winter* rainfall and the amount of salt in the soil and its distribution therein. No field flooded by ordinary sea water this winter is at all likely to bear a crop this year, but if the water is only somewhat brackish, a summer crop is not an impossibility, especially if other factors are favourable. Many cases are known where land flooded by sea water carried quite good crops after one whole winter's rain. On the other hand heavy and poorly drained soils may need five or more years before they can bear crops. Past experience in the coastal parts of Lindsey shows that strong clay land is best left quite untouched for a year and, when work starts, as little weight as possible should be used. Cultivation should be light and begin from the surface. But on the light and well-drained silt lands on which the flooding was of short duration, it may be possible to grow crops this year in Lindsey – and there is little doubt that barley is the most successful crop following flooding by salt water. But it is important to bear in mind that just because a first crop is successful the land has not necessarily recovered. In Holland a second crop has often failed largely because damage to soil structure does not reach the maximum until after the bulk of the salt has been washed out. On some land gypsum treatment may be effective. The gypsum is spread on the surface and is gradually dissolved and washed into the soil. In an average winter little more than a ton of gypsum per acre is likely to be dissolved, but 2–2½ tons per acre should be spread in the dry weather of summer or autumn this year.[1]

These few remarks are sufficient to emphasise that the problem of recovering land from flood water is complicated. Quite a different group of problems arises in considering the feeding of livestock on

[1] The physical properties of clays depend on their chemical composition. In particular, the replacement of calcium, present in normal clays, by sodium from the salt left by flooding with sea water causes the clay to assume intractable properties. The addition of gypsum (calcium sulphate) provides once again an excess of calcium which restores the clay to its original chemical composition and physical properties.

flooded areas. There is no space to develop this subject, and it must suffice to note that some forage crops may be killed, but that some salt-tolerant grasses may survive. Moreover the salty deposits left in the land may be toxic to animals, and some food crops may absorb salt. After effects, such as moulds forming on flooded hay, may all too easily develop. Furthermore the salt content of drinking water used for animals may easily become too high and bring further trouble (Ministry of Agriculture and Fisheries, 1953; Lindsey A.E.C., 1953).

THE SUBSIDENCE OF EASTERN ENGLAND

A subject which has given rise to considerable interest is the downward movement which is known to be taking place in south-eastern England. It is important to emphasise that the movement at the present time can be measured with reasonable accuracy from tide-gauge observations. In the region of the Thames approaches it is of the order of 1–2 mm. a year, which is equivalent to a foot in 150–300 years. This is a significant amount, and throughout a long period of time has made itself felt in various ways. The Thames is now tidal to Teddington; in Roman times it was probably unaffected by tidal influences above London Bridge. The change of level since Roman times may amount to 15 ft, but the subsidence dates back much earlier, and the alternate silt and peat bands that have been so often demonstrated in dock sections imply a sinking which possibly occurred in stages from the time when the North Sea floor was exposed. The earlier stages are the concluding episodes of a long and tangled story dating back to Pliocene times; but the super-position of Roman on Iron and Bronze Age sites at Brentwood gives us a good measure of the prehistoric and historic part of this movement which is still in progress. That the movement may be partly isostatic – the result of subsurface changes resulting from the relief of the weight of the great ice sheets of Pleistocene times – and partly eustatic, in the sense that it is associated with the melting back of the Arctic glaciers today, is discussed in Dr Valentin's account, which also gives references to relevant papers.

FUTURE PROBLEMS

Any long-term planning must, of course, take these isostatic and eustatic movements into consideration. The great flood of January

1953 would have occurred even if this movement had ceased a century ago, but we must realise that two or three centuries ahead the general level of our east coast may well be a foot or more lower than now. This in its turn implies not merely a raising of the level of walls, but also strengthening them so that at their increased height they are still able to hold out the sea. The lessons of the flood also emphasise everywhere that the landward side of the banks needs protection. This is a serious matter, and one that requires much money if it is put into effect. Everywhere on 31 January we heard of the sea overtopping the walls and then scouring out behind them, and in this way producing most of the sea breaches. It may well be that in very vulnerable areas the walls should be duplicated, not so much to prevent entirely an overspill from a very high surge, but in order to have a second line of defence if the outer wall is breached. Whilst it is true that much damage would have been done if the walls had not been breached, it is undoubtedly the case that far more serious flooding resulted from the breaches through which the water rushed.

A study of natural coasts like Scolt Head Island and Blakeney Point demonstrates several points. First, that where beaches and dunes are wide, damage may be small. Next, that where the beach is steep big waves break directly on it and override it and, as opposite Salthouse, may form breaches. Where, too, there is a wide area of sand and salt marsh fronting the coast, little trouble occurred, since the sea spent some of its force well away from the coast. Nevertheless, the unusual height of the surge meant that cliffs and slopes were attacked in unexpected places. The low cliffs near Bacton and Walcott illustrate this point very well. At Winterton, another natural district, the much steeper (i.e. as compared with, say, Scolt Head Island) beach led to a more severe attack on the dunes behind.

All this means that every effort should be made, when it is at all possible, to help the natural accumulation of beach, dunes, shingle ridges and salt marsh. The narrow dunes between Happisburgh and Winterton stood up, but they were sadly weakened. Another storm like that of January 1953 would bring devastation to this stretch of coast and hinterland. It is not easy to alter the beach profiles in this part of Norfolk, nor is there any obvious means of bringing about faster dune growth. The sea wall now being built between Eccles, Horsey and Winterton is the first necessity.

A very similar, but even more urgent problem, faces Lincolnshire. At Sutton and other places the dunes and walls disappeared almost *in toto.* A remarkably good job of new walling and defence has already been done. But if Sutton, Trusthorpe and Mablethorpe are to remain where they are they will need every protection that man and nature can devise. Doubtless the beach will gather again, but can dune growth be engendered? Theddlethorpe and south of Skegness are well protected in this way, because the beach is wider and blowing sand can easily build the dunes. A further trouble in Lincolnshire is the paucity, or even absence, of shingle. There is little or nothing to form a ridge on which dunes might grow. For many years it seems more than probable that the safety of this exposed part of the Lincolnshire coast must depend wholly on man.

Since often the best natural protection is that of a wide beach on which the waves can expend their energy, it follows that it is essential to take an overall picture of the coast. It should be too well known to need emphasis that if the occupiers of a piece of foreshore build groynes to catch the sand and shingle, their neighbours may possibly suffer. Great Yarmouth has undoubtedly gained at the expense of the coast between Gorleston and Lowestoft; Pakefield has lost as a result of Lowestoft harbour works; Southwold harbour holds up material on its north side. These are but examples; many others could be found on our east and south coasts. There may be very good reasons why a particular place should cause beach to accumulate, but are there equally good reasons why, if it so does, another place should suffer? That is the question that has never been satisfactorily answered, and the answer can only be attempted by a central authority trying to see the coast as a unit. Obviously all sorts of problems arise, economic, administrative and others, and to say that what Yarmouth gains is another place's loss is not in itself very helpful since it is difficult to assume that a thriving town could be made to make fundamental alterations in its harbour and promenades.

The loss of life and property at places like Canvey, Jaywick and the central parts of the Lincolshire coasts is to be regretted in every way in so far as the immediate sufferers are concerned. It is, however, right to say that the nation, acting through its central and local government authorities, is in a sense responsible, because if houses and towns are built close behind sea walls and at levels which mean that they could be flooded at any high, or even ordinary, spring tide, those houses are in a potentially dangerous position. Conditions like those existing on 31 January are fortunately rare, but they can and doubtless will recur. They have also occurred in the past. The reasons why the storm of January 1953 will be long remembered are because of the great loss of life and property resulting almost entirely from recent bungalow development and because of its having been so carefully recorded by photography from the air and so fully discussed in the Press. The storm of 1897 and those of earlier times were just as severe, but the number of houses that then existed on reclaimed salt marshes or tidal flats was negligible. Even if this flood had occurred in 1939 the recovery would have been far slower and less effective than it is today. The bulldozer and other war-time and American machines have made a vast difference. Until, however, all houses are out of reach of tidal water they are in potential danger from flooding, although floods of sufficient magnitude may only take place at long intervals of time.

At all places constant care is required to see that sea walls and banks are kept in good condition. Rat holes, drying cracks and untended weaknesses of any sort may lead to disaster. On an open beach where there are dunes every effort should be made to foster their growth. The making of paths through them may, in vulnerable places, lead directly to incursions of sea water. This kind of danger is especially troublesome in places where dune growth is small and where holidaymakers are plentiful. It is difficult to convince people who only visit the coast for a summer holiday that the making of tracks through dunes may be a very real danger (Robinson, 1953). At other places, if a thick growth of vegetation can be encouraged on marshes in front of walls, it may have a beneficial effect. That this is so is evident in parts of Essex. It may not be without significance that some of the major breaks at the northern end of Foulness occurred where the width of marsh in front of the wall was small. Although a good deal is known about the ecology of dunes, shingle and marshes, much more long-term experimental work is required. This can be done at places on the north Norfolk coast where experiments can be allowed to continue unmolested for years if necessary. Nor need these experiments in any way spoil the natural beauty of places like Scolt Head Island and Blakeney Point.

CONCLUSION

The amount of defence work already accomplished since the flooding is impressive and illustrates most effectively what can be done when necessity arises. It was imperative to protect as far as possible existing settlements, since whatever future policies may be adopted it is obvious that, for example, the whole population of Canvey cannot be rehoused in a day! Moreover destruction of dunes and walls has led to the flooding of great areas of good agricultural land, and in the present state of the national economy we must win back that land as soon as possible. The restoration of the *status quo*, however, even if it can be accomplished, is not the only or even the most desirable end. Long-term planning must take much more into consideration: the known subsidence of south-eastern England, the travel of beach material and the effects of groynes and harbour works over long stretches of coast, the growth of dunes and how that growth can be aided or stimulated by human action, the more extensive experimental planting of areas of salt marsh and, still more, of bare sand, as well as the building of protective walls and other defences. It is partly with problems of this type in mind that the Government has appointed a Committee under the Chairmanship of Viscount Waverley. It will be appropriate to end this paper with a statement of the terms of reference of that Committee:

   (i) To examine the causes of the recent floods and the possibilities of a recurrence in Great Britain;

  (ii) To consider what margin of safety for sea defences would be reasonable and practicable having regard on the one hand to the estimated risks involved and on the other to the cost of the protective measures;

 (iii) To consider whether any further measures should be taken by a system of warning or otherwise to lessen the risk of loss of life and serious damage to property.

 (iv) To review the lessons to be learned from the disaster and the administrative and financial responsibilities of the various bodies concerned in providing and maintaining the sea defences and replacing them in the event of damage;
     and to make recommendations.

These terms imply a comprehensive review of the disaster and the lessons to be learned from it. Whilst there is every reason to hope that a storm like that of January 1953 will not recur for many years, there is always the possibility that meterological and tidal conditions may combine to produce a similar crisis any winter. The careful sifting of evidence placed before the Committee, and the appraisal of the points of view of those best able to give advice, take some time. The Government and the country should, however, have a report which will set out clearly the dangers that threaten and the best ways of confronting them.

REFERENCES

BOWDEN, K. F. (1953) 'The peak of a surge of short duration may occur at any state of the tide', *Weather*, LXXXII (Mar).

BRITISH TRANSPORT REVIEW (1953) XI 338.

CORKAN, R. H. (1950) *Phil. Trans. Roy. Soc.*, CCXLII 493; see also *Assoc. Océanogr. Phys. Proc. verbaux*, no. 5 (1962) 167; *Dock and Harbour Authority*, XXVIII (1948) 3.

KING, C. A. M., and BARNES, F. (1953) *Survey* (Nottingham University) III (Mar) 29.

MINISTRY OF AGRICULTURE AND FISHERIES (1953) *Treatment of Land Flooded by Sea Water*, advisory leaflet. See also *Cropping and Treatment of Land which has been Flooded by Salt Water*, supplement to the Lindsey A.E.C. (1953).

RAILWAY MAGAZINE (1953) (May) 301.

ROBINSON, D. N. (1953) 'Trampled dunes may well have helped the attack of the sea on parts of the Lincolnshire coast', *Survey* (Nottingham University) III (Mar) 37.

UFFORD, H. A. Q. V. (1953) 'The onshore direction of the winds on the coast of Holland averaged 50 to 60 knots for 6 to 9 hours before high water', *Weather* (Apr) 116.

Index

This index gives page references to main items. It does not include all place-names mentioned in the text. Many of these are shown in maps or in legends under maps. Those in the index usually refer to areas or lengths of coastline. The names of people listed in the index include those of the writers of the papers in the book and some others to whom special reference is made. The lists of references at the end of each chapter give all relevant information on the papers and books referred to in the text.

1   North coast of Begtrup Vig (aerial photo, 1945)
A recurved spit complex has developed in front of a former bay of the Litorina Sea.
The west part of the shoreline is totally simplified. For explanation of topographical
features and stages of shoreline simplification, cf. Fig. 1.3A.
On the offshore outside the spit complex the white shade indicates sand masses
brought into the bay by beach drifting caused by westerly winds. The bottom in this
shallow-water area shows a surface with runnels and ridges formed by moderate
swell. The east end of this sand platform has a steep slope, the shape of which is due
to current action caused by wind pressure in the narrow opening of the bay between
this sedimentation platform and the corresponding one at the south coast. On these
platforms future bay-closing spit complexes may develop, with an orientation in
continuation of the NW.–SE. running shoreline shown in the left part of the photo.
The seashore limiting line of the field-pattern area indicates the shoreline of the
Litorina Sea.

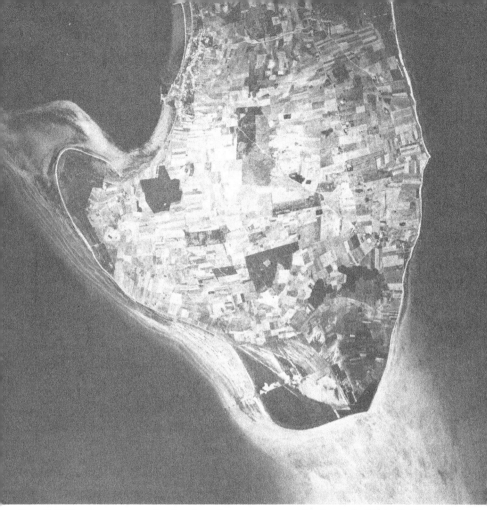

2 *Ahl Hage, a cuspate foreland at the entrance to the bay of Æbeltoft Vig*
  (*cf. Fig. 1.2II*)
  The foreland has developed on a wide platform of sand, Sandhagen, built up by
  beach drifting into the bay. The whole foreland complex acts as a breakwater for the
  harbour of Æbeltoft constructed in the bay behind it. The grain size of the beach
  material diminishes from south to north. The dark shade in the interior of the foreland indicates the conifer plantation established to protect the agricultural area
  against blown sand.

3 Above right: *Spurn Head seen from the south-south-west*
  On the left are the Humber flats, which are covered at high tide, and on the right is
  the North Sea. Despite the shadow cast by the clouds, it is possible to recognise the
  movement of material from north-east to south-west by the lighter shade of grey.

4 Right: *Holderness*

A

B

5   *Cliff sections along the coast
    around Aberystwyth*

A   Boulder clay dipping off 'fossil
    cliff', south-west of Morfa-bychan.

B   'Fossil cliff' running behind
    boulder clay. Note that slopes of
    the old and the Recent cliffs are
    similar. A few yards south of A,
    Morfa-bychan.

C   'Fossil cliff' and overlying boulder
    clay. Slope of old cliff 45°, con-
    trasting with nearly vertical face
    of Recent cliff. South end of
    Aberarth section.

C

D Bevel with boulder-clay platform below, sloping seawards. South of Llanrhystyd.

E Bevel with Recent cliffs below. Between Aberystwyth and Clarach, looking northwards towards Clarach.

F Bevel with freshened cliffs below. In the distance the bevel runs behind boulder clay, which slopes gently to the sea. North end of Aberarth section, looking northwards towards Llanon.

6 *View from Culverhole Point, looking eastwards after the landslip of Christmas 1839*

Pinnacle Rock is in the foreground, with the broke sea cliffs thrust forwards against it, and the newly formed reef of Upper Greensand enclosing a bas of water.
From a lithograph by G. Hawkins, Jr, published by Daniel Dunster (Lyme Regis, 1840).

7 *View of Whitlands Cliff after the landslip of 3 February 1840*

It shows the upper line of inland chalk cliffs, and in front of them the secondary cliff of chalk newly exposed by the subsidence of the seaward part of the undercliff. The cottages have been ruined by the tilting of the ground, which has now formed a hollow flooded by streams diverted by the disturbanc From a lithograph by G. Hawkins, Jr, published by Daniel Dunster (Lyme Regis, 1840).

8 *The great chasm from th fields of Dowlands Farm with Beer Head on the horizon, 7 April 1936*

9 *Ridge of boulder gravel thrown up on the edge of the reef flat at Jabor Island, Jaluit atoll, by typhoon 'Ophelia', 1958*

10 *Close-up of the same ridge, showing imbrication of slab-like boulders of coral, mainly derived from* Acropora *sp.*

11 *Damage to coconut palms on, Mejatto Island, Jaluit atoll, in 1958*

Mejatto is located north of the storm track, and the dominant direction of tree fall is NNW. Note the large number of decapitated palms.

12 *Fresh rubble deposited on an old ground surface on Mejatto Island, Jaluit atoll, 1958*

This photograph is looking SSE., from the direction that the storm winds came.

Printed in the United States
By Bookmasters